R Cookbook

R Cookbook

Paul Teetor

O'REILLY®

Beijing · Cambridge · Farnham · Köln · Sebastopol · Tokyo

R Cookbook
by Paul Teetor

Copyright © 2011 Paul Teetor. All rights reserved.
Printed in the United States of America.

Published by O'Reilly Media, Inc., 1005 Gravenstein Highway North, Sebastopol, CA 95472.

O'Reilly books may be purchased for educational, business, or sales promotional use. Online editions are also available for most titles (*http://my.safaribooksonline.com*). For more information, contact our corporate/institutional sales department: (800) 998-9938 or *corporate@oreilly.com*.

Editor: Mike Loukides	**Indexer:** Jay Marchand
Production Editor: Adam Zaremba	**Cover Designer:** Karen Montgomery
Copyeditor: Matt Darnell	**Interior Designer:** David Futato
Proofreader: Jennifer Knight	**Illustrator:** Robert Romano

Printing History:

March 2011: First Edition.

ISBN: 978-0-596-80915-7

[LSI]

1299082518

Table of Contents

Preface

R is a powerful tool for statistics, graphics, and statistical programming. It is used by tens of thousands of people daily to perform serious statistical analyses. It is a free, open source system whose implementation is the collective accomplishment of many intelligent, hard-working people. There are more than 2,000 available add-ons, and R is a serious rival to all commercial statistical packages.

But R can be frustrating. It's not obvious how to accomplish many tasks, even simple ones. The simple tasks are easy once you know how, yet figuring out that "how" can be maddening.

This book is full of how-to recipes, each of which solves a specific problem. The recipe includes a quick introduction to the solution followed by a discussion that aims to unpack the solution and give you some insight into how it works. I know these recipes are useful and I know they work, because I use them myself.

The range of recipes is broad. It starts with basic tasks before moving on to input and output, general statistics, graphics, and linear regression. Any significant work with R will involve most or all of these areas.

If you are a beginner then this book will get you started faster. If you are an intermediate user, this book is useful for expanding your horizons and jogging your memory ("How do I do that Kolmogorov–Smirnov test again?").

The book is not a tutorial on R, although you will learn something by studying the recipes. It is not a reference manual, but it does contain a lot of useful information. It is not a book on programming in R, although many recipes are useful inside R scripts.

Finally, this book is not an introduction to statistics. Many recipes assume that you are familiar with the underlying statistical procedure, if any, and just want to know how it's done in R.

The Recipes

Most recipes use one or two R functions to solve a specific problem. It's important to remember that I do not describe the functions in detail; rather, I describe just enough to solve the immediate problem. Nearly every such function has additional capabilities beyond those described here, and some have amazing capabilities. I strongly urge you to read the function's help page. You will likely learn something valuable.

Each recipe presents one way to solve a particular problem. Of course, there are likely several reasonable solutions to each problem. When I knew of multiple solutions, I generally selected the simplest one. For any given task, you can probably discover several alternative solutions yourself. This is a cookbook, not a bible.

In particular, R has literally thousands of downloadable add-on packages, many of which implement alternative algorithms and statistical methods. This book concentrates on the core functionality available through the basic distribution, so your best source of alternative solutions may be searching for an add-on package (Recipe 1.11).

A Note on Terminology

The goal of every recipe is to solve a problem and solve it quickly. Rather than laboring in tedious prose, I occasionally streamline the description with terminology that is correct but not precise. A good example is the term "generic function". I refer to print(x) and plot(x) as generic functions because they work for many kinds of x, handling each kind appropriately. A computer scientist would wince at my terminology because, strictly speaking, these are not simply "functions"; they are polymorphic methods with dynamic dispatching. But if I carefully unpacked every such technical detail, the essential solution would be buried in the technicalities. So I just call them functions, which I think is more readable.

Another example, taken from statistics, is the complexity surrounding the semantics of statistical hypothesis testing. Using the strict language of probability theory would obscure the practical application of some tests, so I use more colloquial language when describing each statistical test. See the "Introduction" to Chapter 9 for more about how hypothesis tests are presented in the recipes.

My goal is to make the power of R available to a wide audience by writing readably, not formally. I hope that experts in their respective fields will understand if my terminology is occasionally informal.

Software and Platform Notes

The base distribution of R has frequent and planned releases, but the language definition and core implementation are stable. The recipes in this book should work with any recent release of the base distribution.

Some recipes have platform-specific considerations, and I have carefully noted them. Those recipes mostly deal with software issues, such as installation and configuration. As far as I know, all other recipes will work on all three major platforms for R: Windows, OS X, and Linux/Unix.

Other Resources

On the Web

The mother ship for all things R is the R project site (*http://www.r-project.org*). From there you can download binaries, add-on packages, documentation, and source code as well as many other resources.

Beyond the R project site, I recommend using an R-specific search engine—such as Rseek (*http://rseek.org*), created by Sasha Goodman. You can use a generic search engine, such as Google, but the "R" search term brings up too much extraneous stuff. See Recipe 1.10 for more about searching the Web.

Reading blogs is a great way to learn about R and stay abreast of leading-edge developments. There are surprisingly many such blogs, so I recommend following two blog-of-blogs: R-bloggers (*http://www.r-bloggers.com/*), created by Tal Galili; and PlanetR (*http://planetr.stderr.org/*). By subscribing to their RSS feeds, you will be notified of interesting and useful articles from dozens of websites.

R books

There are many, many books about learning and using R; listed here are a few that I have found useful. Note that the R project site contains an extensive bibliography of books related to R (*http://www.r-project.org/doc/bib/R-books.html*).

I recommend *An Introduction to R*, by William Venables et al. (Network Theory Limited). It covers many topics and is useful for beginners. You can download the PDF for free from CRAN (*http://cran.r-project.org/doc/manuals/R-intro.pdf*); or, better yet, buy the printed copy because the profits are donated to the R project.

R in a Nutshell (*http://oreilly.com/catalog/9780596801717*), by Joseph Adler (O'Reilly), is the quick tutorial and reference you'll keep by your side. It covers many more topics than this *Cookbook*.

Anyone doing serious graphics work in R will want *R Graphics* by Paul Murrell (Chapman & Hall/CRC). Depending on which graphics package you use, you may also want *Lattice: Multivariate Data Visualization with R* by Deepayan Sarkar (Springer) and *ggplot2: Elegant Graphics for Data Analysis* by Hadley Wickham (Springer).

Modern Applied Statistics with S (4th ed.), by William Venables and Brian Ripley (Springer), uses R to illustrate many advanced statistical techniques. The book's functions and datasets are available in the MASS package, which is included in the standard distribution of R.

I'm not wild about any book on programming in R, although new books appear regularly. For programming, I suggest using *R in a Nutshell* together with *S Programming* by William Venables and Brian Ripley (Springer). I also suggest downloading the *R Language Definition (http://cran.r-project.org/doc/manuals/R-lang.pdf)*. The Definition is a work in progress, but it can answer many of your detailed questions regarding R as a programming language.

Statistics books

You will need a good statistics textbook or reference book to accurately interpret the statistical tests performed in R. There are many such fine books—far too many for me to recommend any one above the others.

For learning statistics, a great choice is *Using R for Introductory Statistics* by John Verzani (Chapman & Hall/CRC). It teaches statistics and R together, giving you the necessary computer skills to apply the statistical methods.

Increasingly, statistics authors are using R to illustrate their methods. If you work in a specialized field, then you will likely find a useful and relevant book in the R project bibliography *(http://www.r-project.org/doc/bib/R-books.html)*.

Conventions Used in This Book

The following typographical conventions are used in this book:

Italic

Indicates new terms, URLs, email addresses, filenames, and file extensions.

`Constant width`

Used for program listings, as well as within paragraphs to refer to program elements such as variable or function names, databases, data types, environment variables, statements, and keywords.

`Constant width bold`

Shows commands or other text that should be typed literally by the user.

`Constant width italic`

Shows text that should be replaced with user-supplied values or by values determined by context.

This icon signifies a tip, suggestion, or general note.

This icon indicates a warning or caution.

Using Code Examples

This book is here to help you get your job done. In general, you may use the code in this book in your programs and documentation. You do not need to contact us for permission unless you're reproducing a significant portion of the code. For example, writing a program that uses several chunks of code from this book does not require permission. Selling or distributing a CD-ROM of examples from O'Reilly books does require permission. Answering a question by citing this book and quoting example code does not require permission. Incorporating a significant amount of example code from this book into your product's documentation does require permission.

We appreciate, but do not require, attribution. An attribution usually includes the title, author, publisher, and ISBN. For example: "*R Cookbook* by Paul Teetor. Copyright 2011 Paul Teetor, 978-0-596-80915-7."

If you feel your use of code examples falls outside fair use or the permission just described, feel free to contact us at *permissions@oreilly.com*.

Safari® Books Online

Safari Books Online is an on-demand digital library that lets you easily search over 7,500 technology and creative reference books and videos to find the answers you need quickly.

With a subscription, you can read any page and watch any video from our library online. Read books on your cell phone and mobile devices. Access new titles before they are available for print, get exclusive access to manuscripts in development, and post feedback for the authors. Copy and paste code samples, organize your favorites, download chapters, bookmark key sections, create notes, print out pages, and benefit from many other time-saving features.

O'Reilly Media has uploaded this book to the Safari Books Online service. For full digital access to it and to other books on similar topics from O'Reilly and other publishers, sign up for free at *http://my.safaribooksonline.com*.

How to Contact Us

Please address comments and questions concerning this book to the publisher:

O'Reilly Media, Inc.
1005 Gravenstein Highway North
Sebastopol, CA 95472
800-998-9938 (in the United States or Canada)
707-829-0515 (international or local)
707-829-0104 (fax)

We have a web page for this book, where we list errata, examples, and any additional information. You can access this page at:

http://www.oreilly.com/catalog/9780596809157

To comment or ask technical questions about this book, send email to:

bookquestions@oreilly.com

For more information about our books, courses, conferences, and news, see our website at *http://oreilly.com*.

Find us on Facebook: *http://facebook.com/oreilly*

Follow us on Twitter: *http://twitter.com/oreillymedia*

Watch us on YouTube: *http://www.youtube.com/oreillymedia*

Acknowledgments

With gratitude I thank the R community in general and the R Core Team in particular. Their selfless contributions are enormous. The world of statistics is benefiting tremendously from their work.

I wish to thank the book's technical reviewers: James D. Long, Timothy McMurry, David Reiner, Jeffery Ryan, and John Verzani. My thanks, also, to Joe Adler for his comments on the text. Their feedback was critical for improving the quality, accuracy, and usefulness of this book. They saved me numerous times from showing the world how foolish I really am.

Mike Loukides has been an excellent editor, and I am deeply grateful for his wisdom and guidance. When I started this little project, someone told me that Mike is the best in the business. I believe it.

My greatest gratitude is to my dear wife, Anna. Her support made this book possible. Her partnership made it a joy.

Getting Started and Getting Help

Introduction

This chapter sets the groundwork for the other chapters. It explains how to download, install, and run R.

More importantly, it also explains how to get answers to your questions. The R community provides a wealth of documentation and help. You are not alone. Here are some common sources of help:

Local, installed documentation
> When you install R on your computer, a mass of documentation is also installed. You can browse the local documentation (Recipe 1.6) and search it (Recipe 1.8). I am amazed how often I search the Web for an answer only to discover it was already available in the installed documentation.

Task views (http://cran.r-project.org/web/views)
> A *task view* describes packages that are specific to one area of statistical work, such as econometrics, medical imaging, psychometrics, or spatial statistics. Each task view is written and maintained by an expert in the field. There are 28 such task views, so there is likely to be one or more for your areas of interest. I recommend that every beginner find and read at least one task view in order to gain a sense of R's possibilities (Recipe 1.11).

Package documentation
> Most packages include useful documentation. Many also include overviews and tutorials, called *vignettes* in the R community. The documentation is kept with the packages in package repositories, such as CRAN (*http://cran.r-project.org/*), and it is automatically installed on your machine when you install a package.

Mailing lists
> Volunteers have generously donated many hours of time to answer beginners' questions that are posted to the R mailing lists. The lists are archived, so you can search the archives for answers to your questions (Recipe 1.12).

Question and answer (Q&A) websites

On a Q&A site, anyone can post a question, and knowledgeable people can respond. Readers vote on the answers, so the best answers tend to emerge over time. All this information is tagged and archived for searching. These sites are a cross between a mailing list and a social network; the Stack Overflow site (*http://stacko verflow.com/*) is a good example.

The Web

The Web is loaded with information about R, and there are R-specific tools for searching it (Recipe 1.10). The Web is a moving target, so be on the lookout for new, improved ways to organize and search information regarding R.

1.1 Downloading and Installing R

Problem

You want to install R on your computer.

Solution

Windows and OS X users can download R from CRAN, the Comprehensive R Archive Network. Linux and Unix users can install R packages using their package management tool:

Windows

1. Open *http://www.r-project.org/* in your browser.
2. Click on "CRAN". You'll see a list of mirror sites, organized by country.
3. Select a site near you.
4. Click on "Windows" under "Download and Install R".
5. Click on "base".
6. Click on the link for downloading the latest version of R (an *.exe* file).
7. When the download completes, double-click on the *.exe* file and answer the usual questions.

OS X

1. Open *http://www.r-project.org/* in your browser.
2. Click on "CRAN". You'll see a list of mirror sites, organized by country.
3. Select a site near you.
4. Click on "MacOS X".
5. Click on the *.pkg* file for the latest version of R, under "Files:", to download it.
6. When the download completes, double-click on the *.pkg* file and answer the usual questions.

Linux or Unix

The major Linux distributions have packages for installing R. Here are some examples:

Distribution	Package name
Ubuntu or Debian	r-base
Red Hat or Fedora	R.i386
Suse	R-base

Use the system's package manager to download and install the package. Normally, you will need the root password or sudo privileges; otherwise, ask a system administrator to perform the installation.

Discussion

Installing R on Windows or OS X is straightforward because there are prebuilt binaries for those platforms. You need only follow the preceding instructions. The CRAN Web pages also contain links to installation-related resources, such as frequently asked questions (FAQs) and tips for special situations ("How do I install R when using Windows Vista?") that you may find useful.

Theoretically, you can install R on Linux or Unix in one of two ways: by installing a distribution package or by building it from scratch. In practice, installing a package is the preferred route. The distribution packages greatly streamline both the initial installation and subsequent updates.

On Ubuntu or Debian, use apt-get to download and install R. Run under sudo to have the necessary privileges:

```
$ sudo apt-get install r-base
```

On Red Hat or Fedora, use yum:

```
$ sudo yum install R.i386
```

Most platforms also have graphical package managers, which you might find more convenient.

Beyond the base packages, I recommend installing the documentation packages, too. On my Ubuntu machine, for example, I installed r-base-html (because I like browsing the hyperlinked documentation) as well as r-doc-html, which installs the important R manuals locally:

```
$ sudo apt-get install r-base-html r-doc-html
```

Some Linux repositories also include prebuilt copies of R packages available on CRAN. I don't use them because I'd rather get my software directly from CRAN itself, which usually has the freshest versions.

In rare cases, you may need to build R from scratch. You might have an obscure, un-supported version of Unix; or you might have special considerations regarding per-formance or configuration. The build procedure on Linux or Unix is quite standard. Download the tarball from the home page of your CRAN mirror; it's called something like *R-2.12.1.tar.gz*, except the "2.12.1" will be replaced by the latest version. Unpack the tarball, look for a file called *INSTALL*, and follow the directions.

See Also

R in a Nutshell (O'Reilly) contains more details of downloading and installing R, in-cluding instructions for building the Windows and OS X versions. Perhaps the ultimate guide is the one entitled R Installation and Administration (*http://cran.r-project.org/doc/manuals/R-admin.html*), available on CRAN, which describes building and instal-ling R on a variety of platforms.

This recipe is about installing the base package. See Recipe 3.9 for installing add-on packages from CRAN.

1.2 Starting R

Problem

You want to run R on your computer.

Solution

Windows

Click on Start → All Programs → R; or double-click on the R icon on your desktop (assuming the installer created an icon for you).

OS X

Either click on the icon in the *Applications* directory or put the R icon on the dock and click on the icon there. Alternatively, you can just type R on a Unix command line in a shell.

Linux or Unix

Start the R program from the shell prompt using the R command (uppercase R).

Discussion

How you start R depends upon your platform.

Starting on Windows

When you start R, it opens a new window. The window includes a text pane, called the R Console, where you enter R expressions (see Figure 1-1).

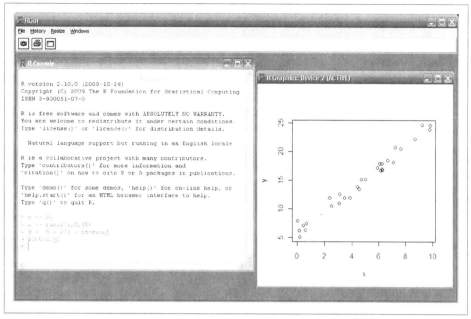

Figure 1-1. R on Windows

There is an odd thing about the Windows Start menu for R. Every time you upgrade to a new version of R, the Start menu expands to contain the new version while keeping all the previously installed versions. So if you've upgraded, you may face several choices such as "R 2.8.1", "R 2.9.1", "R 2.10.1", and so forth. Pick the newest one. (You might also consider uninstalling the older versions to reduce the clutter.)

Using the Start menu is cumbersome, so I suggest starting R in one of two other ways: by creating a desktop shortcut or by double-clicking on your *.RData* file.

The installer may have created a desktop icon. If not, creating a shortcut is easy: follow the Start menu to the R program, but instead of left-clicking to run R, press and hold your mouse's right button on the program name, drag the program name to your desktop, and release the mouse button. Windows will ask if you want to Copy Here or Move Here. Select Copy Here, and the shortcut will appear on your desktop.

Another way to start R is by double-clicking on a *.RData* file in your working directory. This is the file that R creates to save your workspace. The first time you create a directory, start R and change to that directory. Save your workspace there, either by exiting or using the **save.image** function. That will create the *.RData* file. Thereafter, you can simply open the directory in Windows Explorer and then double-click on the *.RData* file to start R.

Perhaps the most baffling aspect of starting R on Windows is embodied in a simple question: When R starts, what is the working directory? The answer, of course, is that "it depends":

- If you start R from the Start menu, the working directory is normally either *C:\Documents and Settings\<username>\My Documents* (Windows XP) or *C:\Users \<username>\Documents* (Windows Vista, Windows 7). You can override this default by setting the R_USER environment variable to an alternative directory path.

- If you start R from a desktop shortcut, you can specify an alternative startup directory that becomes the working directory when R is started. To specify the alternative directory, right-click on the shortcut, select Properties, enter the directory path in the box labeled "Start in", and click OK.

- Starting R by double-clicking on your *.RData* file is the most straightforward solution to this little problem. R will automatically change its working directory to be the file's directory, which is usually what you want.

In any event, you can always use the getwd function to discover your current working directory (Recipe 3.1).

Just for the record, Windows also has a console version of R called *Rterm.exe*. You'll find it in the *bin* subdirectory of your R installation. It is much less convenient than the graphic user interface (GUI) version, and I never use it. I recommend it only for batch (noninteractive) usage such as running jobs from the Windows scheduler. In this book, I assume you are running the GUI version of R, not the console version.

Starting on OS X

Run R by clicking the R icon in the *Applications* folder. (If you use R frequently, you can drag it from the folder to the dock.) That will run the GUI version, which is somewhat more convenient than the console version. The GUI version displays your working directory, which is initially your home directory.

OS X also lets you run the console version of R by typing R at the shell prompt.

Starting on Linux and Unix

Start the console version of R from the Unix shell prompt simply by typing R, the name of the program. Be careful to type an uppercase R, not a lowercase r.

The R program has a bewildering number of command line options. Use the --help option to see the complete list.

See Also

See Recipe 1.4 for exiting from R, Recipe 3.1 for more about the current working directory, Recipe 3.2 for more about saving your workspace, and Recipe 3.11 for suppressing the start-up message. See Chapter 2 of *R in a Nutshell*.

1.3 Entering Commands

Problem

You've started R, and you've got a command prompt. Now what?

Solution

Simply enter expressions at the command prompt. R will evaluate them and print (display) the result. You can use command-line editing to facilitate typing.

Discussion

R prompts you with ">". To get started, just treat R like a big calculator: enter an expression, and R will evaluate the expression and print the result:

```
> 1+1
[1] 2
```

The computer adds one and one, giving two, and displays the result.

The [1] before the 2 might be confusing. To R, the result is a vector, even though it has only one element. R labels the value with [1] to signify that this is the first element of the vector...which is not surprising, since it's the *only* element of the vector.

R will prompt you for input until you type a complete expression. The expression max(1,3,5) is a complete expression, so R stops reading input and evaluates what it's got:

```
> max(1,3,5)
[1] 5
```

In contrast, "max(1,3," is an incomplete expression, so R prompts you for more input. The prompt changes from greater-than (>) to plus (+), letting you know that R expects more:

```
> max(1,3,
+ 5)
[1] 5
```

It's easy to mistype commands, and retyping them is tedious and frustrating. So R includes command-line editing to make life easier. It defines single keystrokes that let you easily recall, correct, and reexecute your commands. My own typical command-line interaction goes like this:

1. I enter an R expression with a typo.
2. R complains about my mistake.
3. I press the up-arrow key to recall my mistaken line.
4. I use the left and right arrow keys to move the cursor back to the error.
5. I use the Delete key to delete the offending characters.

6. I type the corrected characters, which inserts them into the command line.

7. I press Enter to reexecute the corrected command.

That's just the basics. R supports the usual keystrokes for recalling and editing command lines, as listed in Table 1-1.

Table 1-1. Keystrokes for command-line editing

Labeled key	Ctrl-key combination	Effect
Up arrow	Ctrl-P	Recall previous command by moving backward through the history of commands.
Down arrow	Ctrl-N	Move forward through the history of commands.
Backspace	Ctrl-H	Delete the character to the left of cursor.
Delete (Del)	Ctrl-D	Delete the character to the right of cursor.
Home	Ctrl-A	Move cursor to the start of the line.
End	Ctrl-E	Move cursor to the end of the line.
Right arrow	Ctrl-F	Move cursor right (forward) one character.
Left arrow	Ctrl-B	Move cursor left (back) one character.
	Ctrl-K	Delete everything from the cursor position to the end of the line.
	Ctrl-U	Clear the whole darn line and start over.
Tab		Name completion (on some platforms).

On Windows and OS X, you can also use the mouse to highlight commands and then use the usual copy and paste commands to paste text into a new command line.

See Also

See Recipe 2.13. From the Windows main menu, follow Help → Console for a complete list of keystrokes useful for command-line editing.

1.4 Exiting from R

Problem

You want to exit from R.

Solution

Windows
Select File → Exit from the main menu; or click on the red X in the upper-right corner of the window frame.

OS X
> Press CMD-q (apple-q); or click on the red X in the upper-left corner of the window frame.

Linux or Unix
> At the command prompt, press Ctrl-D.

On all platforms, you can also use the q function (as in *quit*) to terminate the program.

```
> q()
```

Note the empty parentheses, which are necessary to call the function.

Discussion

Whenever you exit, R asks if you want to save your workspace. You have three choices:

- Save your workspace and exit.
- Don't save your workspace, but exit anyway.
- Cancel, returning to the command prompt rather than exiting.

If you save your workspace, then R writes it to a file called .RData in the current working directory. This will overwrite the previously saved workspace, if any, so don't save if you don't like the changes to your workspace (e.g., if you have accidentally erased critical data).

See Also

See Recipe 3.1 for more about the current working directory and Recipe 3.2 for more about saving your workspace. See Chapter 2 of *R in a Nutshell*.

1.5 Interrupting R

Problem

You want to interrupt a long-running computation and return to the command prompt without exiting R.

Solution

Windows or OS X
> Either press the Esc key or click on the Stop-sign icon.

Linux or Unix
> Press Ctrl-C. This will interrupt R without terminating it.

Discussion

Interrupting R can leave your variables in an indeterminate state, depending upon how far the computation had progressed. Check your workspace after interrupting.

See Also

See Recipe 1.4.

1.6 Viewing the Supplied Documentation

Problem

You want to read the documentation supplied with R.

Solution

Use the `help.start` function to see the documentation's table of contents:

```
> help.start()
```

From there, links are available to all the installed documentation.

Discussion

The base distribution of R includes a wealth of documentation—literally thousands of pages. When you install additional packages, those packages contain documentation that is also installed on your machine.

It is easy to browse this documentation via the `help.start` function, which opens a window on the top-level table of contents; see Figure 1-2.

The two links in the Reference section are especially useful:

Packages
Click here to see a list of all the installed packages, both in the base packages and the additional, installed packages. Click on a package name to see a list of its functions and datasets.

Search Engine & Keywords
Click here to access a simple search engine, which allows you to search the documentation by keyword or phrase. There is also a list of common keywords, organized by topic; click one to see the associated pages.

See Also

The local documentation is copied from the R Project website (*http://www.r-project .org*), which may have updated documents.

Figure 1-2. Documentation table of contents

1.7 Getting Help on a Function

Problem

You want to know more about a function that is installed on your machine.

Solution

Use help to display the documentation for the function:

```
> help(functionname)
```

Use args for a quick reminder of the function arguments:

```
> args(functionname)
```

Use example to see examples of using the function:

```
> example(functionname)
```

Discussion

I present many R functions in this book. Every R function has more bells and whistles than I can possibly describe. If a function catches your interest, I strongly suggest reading the help page for that function. One of its bells or whistles might be very useful to you.

Suppose you want to know more about the mean function. Use the help function like this:

```
> help(mean)
```

This will either open a window with function documentation or display the documentation on your console, depending upon your platform. A shortcut for the help command is to simply type ? followed by the function name:

```
> ?mean
```

Sometimes you just want a quick reminder of the arguments to a function: What are they, and in what order do they occur? Use the args function:

```
> args(mean)
function (x, ...)
NULL
> args(sd)
function (x, na.rm = FALSE)
NULL
```

The first line of output from args is a synopsis of the function call. For mean, the synopsis shows one argument, x, which is a vector of numbers. For sd, the synopsis shows the same vector, x, and an optional argument called na.rm. (You can ignore the second line of output, which is often just NULL.)

Most documentation for functions includes examples near the end. A cool feature of R is that you can request that it execute the examples, giving you a little demonstration of the function's capabilities. The documentation for the mean function, for instance, contains examples, but you don't need to type them yourself. Just use the example function to watch them run:

```
> example(mean)

mean> x <- c(0:10, 50)

mean> xm <- mean(x)

mean> c(xm, mean(x, trim = 0.1))
[1] 8.75 5.50

mean> mean(USArrests, trim = 0.2)
  Murder  Assault UrbanPop     Rape
    7.42   167.60    66.20    20.16
```

The user typed `example(mean)`. Everything else was produced by R, which executed the examples from the help page and displayed the results.

See Also

See Recipe 1.8 for searching for functions and Recipe 3.5 for more about the search path.

1.8 Searching the Supplied Documentation

Problem

You want to know more about a function that is installed on your machine, but the `help` function reports that it cannot find documentation for any such function.

Alternatively, you want to search the installed documentation for a keyword.

Solution

Use `help.search` to search the R documentation on your computer:

```
> help.search("pattern")
```

A typical *pattern* is a function name or keyword. Notice that it must be enclosed in quotation marks.

For your convenience, you can also invoke a search by using two question marks (in which case the quotes are not required):

```
> ??pattern
```

Discussion

You may occasionally request help on a function only to be told R knows nothing about it:

```
> help(adf.test)
No documentation for 'adf.test' in specified packages and libraries:
you could try 'help.search("adf.test")'
```

This can be frustrating if you *know* the function is installed on your machine. Here the problem is that the function's package is not currently loaded, and you don't know which package contains the function. It's a kind of catch-22 (the error message indicates the package is not currently in your search path, so R cannot find the help file; see Recipe 3.5 for more details).

The solution is to search all your installed packages for the function. Just use the `help.search` function, as suggested in the error message:

```
> help.search("adf.test")
```

The search will produce a listing of all packages that contain the function:

```
Help files with alias or concept or title matching 'adf.test' using
regular expression matching:

tseries::adf.test        Augmented Dickey-Fuller Test

Type '?PKG::FOO' to inspect entry 'PKG::FOO TITLE'.
```

The following output, for example, indicates that the tseries package contains the adf.test function. You can see its documentation by explicitly telling help which package contains the function:

```
> help(adf.test, package="tseries")
```

Alternatively, you can insert the tseries package into your search list and repeat the original help command, which will then find the function and display the documentation.

You can broaden your search by using keywords. R will then find any installed documentation that contains the keywords. Suppose you want to find all functions that mention the Augmented Dickey–Fuller (ADF) test. You could search on a likely pattern:

```
> help.search("dickey-fuller")
```

On my machine, the result looks like this because I've installed two additional packages (fUnitRoots and urca) that implement the ADF test:

```
Help files with alias or concept or title matching 'dickey-fuller' using
fuzzy matching:

fUnitRoots::DickeyFullerPValues
                         Dickey-Fuller p Values
tseries::adf.test        Augmented Dickey-Fuller Test
urca::ur.df              Augmented-Dickey-Fuller Unit Root Test

Type '?PKG::FOO' to inspect entry 'PKG::FOO TITLE'.
```

See Also

You can also access the local search engine through the documentation browser; see Recipe 1.6 for how this is done. See Recipe 3.5 for more about the search path and Recipe 4.4 for getting help on functions.

1.9 Getting Help on a Package

Problem

You want to learn more about a package installed on your computer.

Solution

Use the `help` function and specify a package name (without a function name):

```
> help(package="packagename")
```

Discussion

Sometimes you want to know the contents of a package (the functions and datasets). This is especially true after you download and install a new package, for example. The help function can provide the contents plus other information once you specify the package name.

This call to help will display the information for the `tseries` package, a standard package in the base distribution:

```
> help(package="tseries")
```

The information begins with a description and continues with an index of functions and datasets. On my machine, the first few lines look like this:

```
            Information on package 'tseries'

Description:

Package:          tseries
Version:          0.10-22
Date:             2009-11-22
Title:            Time series analysis and computational finance
Author:           Compiled by Adrian Trapletti
                  <a.trapletti@swissonline.ch>
Maintainer:       Kurt Hornik <Kurt.Hornik@R-project.org>
Description:      Package for time series analysis and computational
                  finance
Depends:          R (>= 2.4.0), quadprog, stats, zoo
Suggests:         its
Imports:          graphics, stats, utils
License:          GPL-2
Packaged:         2009-11-22 19:03:45 UTC; hornik
Repository:       CRAN
Date/Publication: 2009-11-22 19:06:50
Built:            R 2.10.0; i386-pc-mingw32; 2009-12-01 19:32:47 UTC;
                  windows

Index:

NelPlo            Nelson-Plosser Macroeconomic Time Series
USeconomic        U.S. Economic Variables
adf.test          Augmented Dickey-Fuller Test
arma              Fit ARMA Models to Time Series

.
. (etc.)
.
```

Some packages also include vignettes, which are additional documents such as introductions, tutorials, or reference cards. They are installed on your computer as part of the package documentation when you install the package. The help page for a package includes a list of its vignettes near the bottom.

You can see a list of all vignettes on your computer by using the `vignette` function:

```
> vignette()
```

You can see the vignettes for a particular package by including its name:

```
> vignette(package="packagename")
```

Each vignette has a name, which you use to view the vignette:

```
> vignette("vignettename")
```

See Also

See Recipe 1.7 for getting help on a particular function in a package.

1.10 Searching the Web for Help

Problem

You want to search the Web for information and answers regarding R.

Solution

Inside R, use the `RSiteSearch` function to search by keyword or phrase:

```
> RSiteSearch("key phrase")
```

Inside your browser, try using these sites for searching:

http://rseek.org
> This is a Google custom search that is focused on R-specific websites.

http://stackoverflow.com/
> Stack Overflow is a searchable Q&A site oriented toward programming issues such as data structures, coding, and graphics.

http://stats.stackexchange.com/
> The Statistical Analysis area on Stack Exchange is also a searchable Q&A site, but it is oriented more toward statistics than programming.

Discussion

The `RSiteSearch` function will open a browser window and direct it to the search engine on the R Project website (*http://search.r-project.org/*). There you will see an initial search that you can refine. For example, this call would start a search for "canonical correlation":

```
> RSiteSearch("canonical correlation")
```

This is quite handy for doing quick web searches without leaving R. However, the search scope is limited to R documentation and the mailing-list archives.

The *rseek.org* site provides a wider search. Its virtue is that it harnesses the power of the Google search engine while focusing on sites relevant to R. That eliminates the extraneous results of a generic Google search. The beauty of *rseek.org* is that it organizes the results in a useful way.

Figure 1-3 shows the results of visiting *rseek.org* and searching for "canonical correlation". The left side of the page shows general results for search R sites. The right side is a tabbed display that organizes the search results into several categories:

- Introductions
- Task Views
- Support Lists
- Functions
- Books
- Blogs
- Related Tools

Figure 1-3. Search results from rseek.org

If you click on the Introductions tab, for example, you'll find tutorial material. The Task Views tab will show any Task View that mentions your search term. Likewise, clicking on Functions will show links to relevant R functions. This is a good way to zero in on search results.

Stack Overflow (*http://stackoverflow.com/*) is a so-called Q&A site, which means that anyone can submit a question and experienced users will supply answers—often there are multiple answers to each question. Readers vote on the answers, so good answers tend to rise to the top. This creates a rich database of Q&A dialogs, which you can search. Stack Overflow is strongly problem oriented, and the topics lean toward the programming side of R.

Stack Overflow hosts questions for many programming languages; therefore, when entering a term into their search box, prefix it with "[r]" to focus the search on questions tagged for R. For example, searching via "[r] standard error" will select only the questions tagged for R and will avoid the Python and C++ questions.

Stack Exchange (not Overflow) has a Q&A area for Statistical Analysis. The area is more focused on statistics than programming, so use this site when seeking answers that are more concerned with statistics in general and less with R in particular.

See Also

If your search reveals a useful package, use Recipe 3.9 to install it on your machine.

1.11 Finding Relevant Functions and Packages

Problem

Of the 2,000+ packages for R, you have no idea which ones would be useful to you.

Solution

- Visit the list of task views at *http://cran.r-project.org/web/views/*. Find and read the task view for your area, which will give you links to and descriptions of relevant packages. Or visit *http://rseek.org*, search by keyword, click on the Task Views tab, and select an applicable task view.
- Visit crantastic (*http://crantastic.org/*) and search for packages by keyword.
- To find relevant functions, visit *http://rseek.org*, search by name or keyword, and click on the Functions tab.

Discussion

This problem is especially vexing for beginners. You think R can solve your problems, but you have no idea which packages and functions would be useful. A common

question on the mailing lists is: "Is there a package to solve problem X?" That is the silent scream of someone drowning in R.

As of this writing, there are more than 2,000 packages available for free download from CRAN. Each package has a summary page with a short description and links to the package documentation. Once you've located a potentially interesting package, you would typically click on the "Reference manual" link to view the PDF documentation with full details. (The summary page also contains download links for installing the package, but you'll rarely install the package that way; see Recipe 3.9.)

Sometimes you simply have a generic interest—such as Bayesian analysis, econometrics, optimization, or graphics. CRAN contains a set of *task view* pages describing packages that may be useful. A task view is a great place to start since you get an overview of what's available. You can see the list of task view pages at *http://cran.r -project.org/web/views/* or search for them as described in the Solution.

Suppose you happen to know the name of a useful package—say, by seeing it mentioned online. A complete, alphabetical list of packages is available at *http://cran.r -project.org/web/packages/* with links to the package summary pages.

See Also

You can download and install an R package called sos that provides powerful other ways to search for packages; see the vignette at *http://cran.r-project.org/web/packages/ sos/vignettes/sos.pdf*.

1.12 Searching the Mailing Lists

Problem

You have a question, and you want to search the archives of the mailing lists to see whether your question was answered previously.

Solution

- Open *http://rseek.org* in your browser. Search for a keyword or other search term from your question. When the search results appear, click on the "Support Lists" tab.
- You can perform a search within R itself. Use the RSiteSearch function to initiate a search:

    ```
    > RSiteSearch("keyphrase")
    ```

The initial search results will appear in a browser. Under "Target", select the R-help sources, clear the other sources, and resubmit your query.

Discussion

This recipe is really just an application of Recipe 1.10. But it's an important application because you should search the mailing list archives before submitting a new question to the list. Your question has probably been answered before.

See Also

CRAN has a list of additional resources for searching the Web; see *http://cran.r-project .org/search.html*.

1.13 Submitting Questions to the Mailing Lists

Problem

You want to submit a question to the R community via the R-help mailing list.

Solution

The Mailing Lists (*http://www.r-project.org/mail.html*) page contains general information and instructions for using the R-help mailing list. Here is the general process:

1. Subscribe to the R-help list at the Main R Mailing List (*https://stat.ethz.ch/mailman/ listinfo/r-help*).
2. Read the Posting Guide (*http://www.r-project.org/posting-guide.html*) for instructions on writing an effective submission.
3. Write your question carefully and correctly. If appropriate, include a minimal self-reproducing example so that others can reproduce your error or problem.
4. Mail your question to *r-help@r-project.org*.

Discussion

The R mailing list is a powerful resource, but please treat it as a last resort. Read the help pages, read the documentation, search the help list archives, and search the Web. It is most likely that your question has already been answered. Don't kid yourself: very few questions are unique.

After writing your question, submitting it is easy. Just mail it to *r-help@r-project.org*. You must be a list subscriber, however; otherwise your email submission may be rejected.

Your question might arise because your R code is causing an error or giving unexpected results. In that case, a critical element of your question is the *minimal self-contained example*:

Minimal

> Construct the smallest snippet of R code that displays your problem. Remove everything that is irrelevant.

Self-contained

> Include the data necessary to exactly reproduce the error. If the list readers can't reproduce it, they can't diagnose it. For complicated data structures, use the dump function to create an ASCII representation of your data and include it in your message.

Including an example clarifies your question and greatly increases the probability of getting a useful answer.

There are actually several mailing lists. R-help is the main list for general questions. There are also many special interest group (SIG) mailing lists dedicated to particular domains such as genetics, finance, R development, and even R jobs. You can see the full list at *https://stat.ethz.ch/mailman/listinfo*. If your question is specific to one such domain, you'll get a better answer by selecting the appropriate list. As with R-help, however, carefully search the SIG list archives before submitting your question.

See Also

An excellent essay by Eric Raymond and Rick Moen is entitled "How to Ask Questions the Smart Way" (*http://www.catb.org/~esr/faqs/smart-questions.html*). I suggest that you read it before submitting any question.

Some Basics

Introduction

The recipes in this chapter lie somewhere between problem-solving ideas and tutorials. Yes, they solve common problems, but the Solutions showcase common techniques and idioms used in most R code, including the code in this Cookbook. If you are new to R, I suggest skimming this chapter to acquaint yourself with these idioms.

2.1 Printing Something

Problem

You want to display the value of a variable or expression.

Solution

If you simply enter the variable name or expression at the command prompt, R will print its value. Use the `print` function for generic printing of any object. Use the `cat` function for producing custom formatted output.

Discussion

It's very easy to ask R to print something: just enter it at the command prompt:

```
> pi
[1] 3.141593
> sqrt(2)
[1] 1.414214
```

When you enter expressions like that, R evaluates the expression and then implicitly calls the `print` function. So the previous example is identical to this:

```
> print(pi)
[1] 3.141593
> print(sqrt(2))
[1] 1.414214
```

The beauty of print is that it knows how to format any R value for printing, including structured values such as matrices and lists:

```
> print(matrix(c(1,2,3,4), 2, 2))
     [,1] [,2]
[1,]    1    3
[2,]    2    4
> print(list("a","b","c"))
[[1]]
[1] "a"

[[2]]
[1] "b"

[[3]]
[1] "c"
```

This is useful because you can always view your data: just print it. You needn't write special printing logic, even for complicated data structures.

The print function has a significant limitation, however: it prints only one object at a time. Trying to print multiple items gives this mind-numbing error message:

```
> print("The zero occurs at", 2*pi, "radians.")
Error in print.default("The zero occurs at", 2 * pi, "radians.") :
        unimplemented type 'character' in 'asLogical'
```

The only way to print multiple items is to print them one at a time, which probably isn't what you want:

```
> print("The zero occurs at"); print(2*pi); print("radians")
[1] "The zero occurs at"
[1] 6.283185
[1] "radians"
```

The cat function is an alternative to print that lets you combine multiple items into a continuous output:

```
> cat("The zero occurs at", 2*pi, "radians.", "\n")
The zero occurs at 6.283185 radians.
```

Notice that cat puts a space between each item by default. You must provide a newline character (\n) to terminate the line.

The cat function can print simple vectors, too:

```
> fib <- c(0,1,1,2,3,5,8,13,21,34)
> cat("The first few Fibonacci numbers are:", fib, "...\n")
The first few Fibonacci numbers are: 0 1 1 2 3 5 8 13 21 34 ...
```

Using cat gives you more control over your output, which makes it especially useful in R scripts. A serious limitation, however, is that it cannot print compound data structures such as matrices and lists. Trying to cat them only produces another mind-numbing message:

```
> cat(list("a","b","c"))
Error in cat(list(...), file, sep, fill, labels, append) :
        argument 1 (type 'list') cannot be handled by 'cat'
```

See Also

See Recipe 4.2 for controlling output format.

2.2 Setting Variables

Problem

You want to save a value in a variable.

Solution

Use the assignment operator (<-). There is no need to declare your variable first:

```
> x <- 3
```

Discussion

Using R in "calculator mode" gets old pretty fast. Soon you will want to define variables and save values in them. This reduces typing, saves time, and clarifies your work.

There is no need to declare or explicitly create variables in R. Just assign a value to the name and R will create the variable:

```
> x <- 3
> y <- 4
> z <- sqrt(x^2 + y^2)
> print(z)
[1] 5
```

Notice that the assignment operator is formed from a less-than character (<) and a hyphen (-) with no space between them.

When you define a variable at the command prompt like this, the variable is held in your *workspace*. The workspace is held in the computer's main memory but can be saved to disk when you exit from R. The variable definition remains in the workspace until you remove it.

R is a *dynamically typed language*, which means that we can change a variable's data type at will. We could set x to be numeric, as just shown, and then turn around and immediately overwrite that with (say) a vector of character strings. R will not complain:

```
> x <- 3
> print(x)
[1] 3
> x <- c("fee", "fie", "foe", "fum")
> print(x)
[1] "fee" "fie" "foe" "fum"
```

In some R functions you will see assignment statements that use the strange-looking assignment operator <<-:

```
x <<- 3
```

That forces the assignment to a global variable rather than a local variable.

In the spirit of full disclosure, I will reveal that R also supports two other forms of assignment statements. A single equal sign (=) can be used as an assignment operator at the command prompt. A rightward assignment operator (->) can be used anywhere the leftward assignment operator (<-) can be used:

```
> foo = 3
> print(foo)
[1] 3
> 5 -> fum
> print(fum)
[1] 5
```

These forms are never used in this book, and I recommend that you avoid them. The equals-sign assignment is easily confused with the test for equality. The rightward assignment is just too unconventional and, worse, becomes difficult to read when the expression is long.

See Also

See Recipes 2.4, 2.14, and 3.2. See also the help page for the `assign` function.

2.3 Listing Variables

Problem

You want to know what variables and functions are defined in your workspace.

Solution

Use the `ls` function. Use `ls.str` for more details about each variable.

Discussion

The `ls` function displays the names of objects in your workspace:

```
> x <- 10
> y <- 50
> z <- c("three", "blind", "mice")
```

```
> f <- function(n,p) sqrt(p*(1-p)/n)
> ls()
[1] "f" "x" "y" "z"
```

Notice that ls returns a vector of character strings in which each string is the name of one variable or function. When your workspace is empty, ls returns an empty vector, which produces this puzzling output:

```
> ls()
character(0)
```

That is R's quaint way of saying that ls returned a zero-length vector of strings; that is, it returned an empty vector because nothing is defined in your workspace.

If you want more than just a list of names, try ls.str; this will also tell you something about each variable:

```
> ls.str()
f :  function (n, p)
x :  num 10
y :  num 50
z :  chr [1:3] "three" "blind" "mice"
```

The function is called ls.str because it is both listing your variables and applying the str function to them, showing their structure (Recipe 12.15).

Ordinarily, ls does not return any name that begins with a dot (.). Such names are considered hidden and are not normally of interest to users. (This mirrors the Unix convention of not listing files whose names begin with dot.) You can force ls to list everything by setting the all.names argument to TRUE:

```
> .hidvar <- 10
> ls()
[1] "f" "x" "y" "z"
> ls(all.names=TRUE)
[1] ".hidvar" "f"       "x"       "y"       "z"
```

See Also

See Recipe 2.4 for deleting variables and Recipe 12.15 for inspecting your variables.

2.4 Deleting Variables

Problem

You want to remove unneeded variables or functions from your workspace or to erase its contents completely.

Solution

Use the rm function.

Discussion

Your workspace can get cluttered quickly. The rm function removes, permanently, one or more objects from the workspace:

```
> x <- 2*pi
> x
[1] 6.283185
> rm(x)
> x
Error: object "x" not found
```

There is no "undo"; once the variable is gone, it's gone.

You can remove several variables at once:

```
> rm(x,y,z)
```

You can even erase your entire workspace at once. The rm function has a list argument consisting of a vector of names of variables to remove. Recall that the ls function returns a vector of variables names; hence you can combine rm and ls to erase everything:

```
> ls()
[1] "f" "x" "y" "z"
> rm(list=ls())
> ls()
character(0)
```

Be Polite

Never put rm(list=ls()) into code you share with others, such as a library function or sample code sent to a mailing list. Deleting all the variables in someone else's workspace is worse than rude and will make you extremely unpopular.

See Also

See Recipe 2.3.

2.5 Creating a Vector

Problem

You want to create a vector.

Solution

Use the c(...) operator to construct a vector from given values.

Discussion

Vectors are a central component of R, not just another data structure. A vector can contain either numbers, strings, or logical values but not a mixture.

The c(...) operator can construct a vector from simple elements:

```
> c(1,1,2,3,5,8,13,21)
[1]  1  1  2  3  5  8 13 21
> c(1*pi, 2*pi, 3*pi, 4*pi)
[1]  3.141593  6.283185  9.424778 12.566371
> c("Everyone", "loves", "stats.")
[1] "Everyone" "loves"    "stats."
> c(TRUE,TRUE,FALSE,TRUE)
[1]  TRUE  TRUE FALSE  TRUE
```

If the arguments to c(...) are themselves vectors, it flattens them and combines them into one single vector:

```
> v1 <- c(1,2,3)
> v2 <- c(4,5,6)
> c(v1,v2)
[1] 1 2 3 4 5 6
```

Vectors cannot contain a mix of data types, such as numbers and strings. If you create a vector from mixed elements, R will try to accommodate you by converting one of them:

```
> v1 <- c(1,2,3)
> v3 <- c("A","B","C")
> c(v1,v3)
[1] "1" "2" "3" "A" "B" "C"
```

Here, the user tried to create a vector from both numbers and strings. R converted all the numbers to strings before creating the vector, thereby making the data elements compatible.

Technically speaking, two data elements can coexist in a vector only if they have the same *mode*. The modes of 3.1415 and "foo" are numeric and character, respectively:

```
> mode(3.1415)
[1] "numeric"
> mode("foo")
[1] "character"
```

Those modes are incompatible. To make a vector from them, R converts 3.1415 to character mode so it will be compatible with "foo":

```
> c(3.1415, "foo")
[1] "3.1415" "foo"
> mode(c(3.1415, "foo"))
[1] "character"
```

 c is a generic operator, which means that it works with many datatypes and not just vectors. However, it might not do exactly what you expect, so check its behavior before applying it to other datatypes and objects.

See Also

See the "Introduction" to Chapter 5 for more about vectors and other data structures.

2.6 Computing Basic Statistics

Problem

You want to calculate basic statistics: mean, median, standard deviation, variance, correlation, or covariance.

Solution

Use one of these functions as applies, assuming that x and y are vectors:

- mean(x)
- median(x)
- sd(x)
- var(x)
- cor(x, y)
- cov(x, y)

Discussion

When I first opened the documentation for R, I begin searching for material entitled "Procedures for Calculating Standard Deviation." I figured that such an important topic would likely require a whole chapter.

It's not that complicated.

Standard deviation and other basic statistics are calculated by simple functions. Ordinarily, the function argument is a vector of numbers and the function returns the calculated statistic:

```
> x <- c(0,1,1,2,3,5,8,13,21,34)
> mean(x)
[1] 8.8
> median(x)
[1] 4
> sd(x)
[1] 11.03328
> var(x)
[1] 121.7333
```

The `sd` function calculates the sample standard deviation, and `var` calculates the sample variance.

The `cor` and `cov` functions can calculate the correlation and covariance, respectively, between two vectors:

```
> x <- c(0,1,1,2,3,5,8,13,21,34)
> y <- log(x+1)
> cor(x,y)
[1] 0.9068053
> cov(x,y)
[1] 11.49988
```

All these functions are picky about values that are not available (NA). Even one NA value in the vector argument causes any of these functions to return NA or even halt altogether with a cryptic error:

```
> x <- c(0,1,1,2,3,NA)
> mean(x)
[1] NA
> sd(x)
[1] NA
```

It's annoying when R is that cautious, but it is the right thing to do. You must think carefully about your situation. Does an NA in your data invalidate the statistic? If yes, then R is doing the right thing. If not, you can override this behavior by setting `na.rm=TRUE`, which tells R to ignore the NA values:

```
> x <- c(0,1,1,2,3,NA)
> mean(x, na.rm=TRUE)
[1] 1.4
> sd(x, na.rm=TRUE)
[1] 1.140175
```

A beautiful aspect of `mean` and `sd` is that they are smart about data frames. They understand that each column of the data frame is a different variable, so they calculate their statistic for each column individually. This example calculates those basic statistics for a data frame with three columns:

```
> print(dframe)
        small    medium       big
1   0.6739635 10.526448  99.83624
2   1.5524619  9.205156 100.70852
3   0.3250562 11.427756  99.73202
4   1.2143595  8.533180  98.53608
5   1.3107692  9.763317 100.74444
6   2.1739663  9.806662  98.58961
7   1.6187899  9.150245 100.46707
8   0.8872657 10.058465  99.88068
9   1.9170283  9.182330 100.46724
10 0.7767406  7.949692 100.49814
> mean(dframe)
    small    medium       big
 1.245040  9.560325 99.946003
> sd(dframe)
```

```
      small    medium       big
 0.5844025 0.9920281 0.8135498
```

Notice that mean and sd both return three values, one for each column defined by the data frame. (Technically, they return a three-element vector whose names attribute is taken from the columns of the data frame.)

The var function understands data frames, too, but it behaves quite differently than do mean and sd. It calculates the covariance between the columns of the data frame and returns the covariance matrix:

```
> var(dframe)
               small       medium         big
small     0.34152627 -0.21516416 -0.04005275
medium   -0.21516416  0.98411974 -0.09253855
big      -0.04005275 -0.09253855  0.66186326
```

Likewise, if x is either a data frame or a matrix, then cor(x) returns the correlation matrix and cov(x) returns the covariance matrix:

```
> cor(dframe)
               small       medium         big
small     1.00000000 -0.3711367 -0.08424345
medium   -0.37113670  1.0000000 -0.11466070
big      -0.08424345 -0.1146607  1.00000000
> cov(dframe)
               small       medium         big
small     0.34152627 -0.21516416 -0.04005275
medium   -0.21516416  0.98411974 -0.09253855
big      -0.04005275 -0.09253855  0.66186326
```

The median function does not understand data frames, alas. To calculate the medians of data frame columns, use Recipe 6.4 and apply the median function to each column separately.

See Also

See Recipes 2.14, 6.4, and 9.17.

2.7 Creating Sequences

Problem

You want to create a sequence of numbers.

Solution

Use an n:m expression to create the simple sequence n, n+1, n+2, ..., m:

```
> 1:5
[1] 1 2 3 4 5
```

Use the `seq` function for sequences with an increment other than 1:

```
> seq(from=1, to=5, by=2)
[1] 1 3 5
```

Use the `rep` function to create a series of repeated values:

```
> rep(1, times=5)
[1] 1 1 1 1 1
```

Discussion

The colon operator (`n:m`) creates a vector containing the sequence n, $n+1$, $n+2$, ..., m:

```
> 0:9
[1] 0 1 2 3 4 5 6 7 8 9
> 10:19
[1] 10 11 12 13 14 15 16 17 18 19
> 9:0
[1] 9 8 7 6 5 4 3 2 1 0
```

Observe that R was clever with the last expression (`9:0`). Because 9 is larger than 0, it counts backward from the starting to ending value.

The colon operator works for sequences that grow by 1 only. The `seq` function also builds sequences but supports an optional third argument, which is the increment:

```
> seq(from=0, to=20)
 [1]  0  1  2  3  4  5  6  7  8  9 10 11 12 13 14 15 16 17 18 19 20
> seq(from=0, to=20, by=2)
 [1]  0  2  4  6  8 10 12 14 16 18 20
> seq(from=0, to=20, by=5)
[1]  0  5 10 15 20
```

Alternatively, you can specify a length for the output sequence and then R will calculate the necessary increment:

```
> seq(from=0, to=20, length.out=5)
[1]  0  5 10 15 20
> seq(from=0, to=100, length.out=5)
[1]   0  25  50  75 100
```

The increment need not be an integer. R can create sequences with fractional increments, too:

```
> seq(from=1.0, to=2.0, length.out=5)
[1] 1.00 1.25 1.50 1.75 2.00
```

For the special case of a "sequence" that is simply a repeated value you should use the `rep` function, which repeats its first argument:

```
> rep(pi, times=5)
[1] 3.141593 3.141593 3.141593 3.141593 3.141593
```

See Also

See Recipe 7.14 for creating a sequence of Date objects.

2.8 Comparing Vectors

Problem

You want to compare two vectors or you want to compare an entire vector against a scalar.

Solution

The comparison operators (==, !=, <, >, <=, >=) can perform an element-by-element comparison of two vectors. They can also compare a vector's element against a scalar. The result is a vector of logical values in which each value is the result of one element-wise comparison.

Discussion

R has two *logical values*, TRUE and FALSE. These are often called *Boolean values* in other programming languages.

The comparison operators compare two values and return TRUE or FALSE, depending upon the result of the comparison:

```
> a <- 3
> a == pi     # Test for equality
[1] FALSE
> a != pi     # Test for inequality
[1] TRUE
> a < pi
[1] TRUE
> a > pi
[1] FALSE
> a <= pi
[1] TRUE
> a >= pi
[1] FALSE
```

You can experience the power of R by comparing entire vectors at once. R will perform an element-by-element comparison and return a vector of logical values, one for each comparison:

```
> v <- c( 3, pi,  4)
> w <- c(pi, pi, pi)
> v == w                 # Compare two 3-element vectors
[1] FALSE  TRUE FALSE    # Result is a 3-element vector
> v != w
[1]  TRUE FALSE  TRUE
> v < w
[1]  TRUE FALSE FALSE
> v <= w
[1]  TRUE  TRUE FALSE
> v > w
[1] FALSE FALSE  TRUE
```

```
> v >= w
[1] FALSE  TRUE  TRUE
```

You can also compare a vector against a single scalar, in which case R will expand the scalar to the vector's length and then perform the element-wise comparison. The previous example can be simplified in this way:

```
> v <- c(3, pi, 4)
> v == pi              # Compare a 3-element vector against one number
[1] FALSE  TRUE FALSE
> v != pi
[1]  TRUE FALSE  TRUE
```

(This is an application of the Recycling Rule, Recipe 5.3.)

After comparing two vectors, you often want to know whether *any* of the comparisons were true or whether *all* the comparisons were true. The any and all functions handle those tests. They both test a logical vector. The any function returns TRUE if any element of the vector is TRUE. The all function returns TRUE if all elements of the vector are TRUE:

```
> v <- c(3, pi, 4)
> any(v == pi)         # Return TRUE if any element of v equals pi
[1] TRUE
> all(v == 0)          # Return TRUE if all elements of v are zero
[1] FALSE
```

See Also

See Recipe 2.9.

2.9 Selecting Vector Elements

Problem

You want to extract one or more elements from a vector.

Solution

Select the indexing technique appropriate for your problem:

- Use square brackets to select vector elements by their position, such as v[3] for the third element of v.
- Use negative indexes to exclude elements.
- Use a vector of indexes to select multiple values.
- Use a logical vector to select elements based on a condition.
- Use names to access named elements.

Discussion

Selecting elements from vectors is another powerful feature of R. Basic selection is handled just as in many other programming languages—use square brackets and a simple index:

```
> fib <- c(0,1,1,2,3,5,8,13,21,34)
> fib
 [1]  0  1  1  2  3  5  8 13 21 34
> fib[1]
[1] 0
> fib[2]
[1] 1
> fib[3]
[1] 1
> fib[4]
[1] 2
> fib[5]
[1] 3
```

Notice that the first element has an index of 1, not 0 as in some other programming languages.

A cool feature of vector indexing is that you can select multiple elements at once. The index itself can be a vector, and each element of that indexing vector selects an element from the data vector:

```
> fib[1:3]        # Select elements 1 through 3
[1] 0 1 1
> fib[4:9]        # Select elements 4 through 9
[1]  2  3  5  8 13 21
```

An index of 1:3 means select elements 1, 2, and 3, as just shown. The indexing vector needn't be a simple sequence, however. You can select elements anywhere within the data vector—as in this example, which selects elements 1, 2, 4, and 8:

```
> fib[c(1,2,4,8)]
[1]  0  1  2 13
```

R interprets negative indexes to mean *exclude* a value. An index of –1, for instance, means exclude the first value and return all other values:

```
> fib[-1]         # Ignore first element
[1]  1  1  2  3  5  8 13 21 34
```

This method can be extended to exclude whole slices by using an indexing vector of negative indexes:

```
> fib[1:3]        # As before
[1] 0 1 1
> fib[-(1:3)]           # Invert sign of index to exclude instead of select
[1]  2  3  5  8 13 21 34
```

Another indexing technique uses a logical vector to select elements from the data vector. Everywhere that the logical vector is TRUE, an element is selected:

```
> fib < 10                    # This vector is TRUE wherever fib is less than 10
 [1]  TRUE  TRUE  TRUE  TRUE  TRUE  TRUE  TRUE FALSE FALSE FALSE
> fib[fib < 10]               # Use that vector to select elements less than 10
 [1] 0 1 1 2 3 5 8

> fib %% 2 == 0               # This vector is TRUE wherever fib is even
 [1]  TRUE FALSE FALSE  TRUE FALSE FALSE  TRUE FALSE FALSE  TRUE
> fib[fib %% 2 == 0]          # Use that vector to select the even elements
 [1]  0  2  8 34
```

Ordinarily, the logical vector should be the same length as the data vector so you are clearly either including or excluding each element. (If the lengths differ then you need to understand the Recycling Rule, Recipe 5.3.)

By combining vector comparisons, logical operators, and vector indexing, you can perform powerful selections with very little R code:

Select all elements greater than the median

```
v[ v > median(v) ]
```

Select all elements in the lower and upper 5%

```
v[ (v < quantile(v,0.05)) | (v > quantile(v,0.95)) ]
```

Select all elements that exceed ±2 standard deviations from the mean

```
v[ abs(v-mean(v)) > 2*sd(v) ]
```

Select all elements that are neither NA nor NULL

```
v[ !is.na(v) & !is.null(v) ]
```

One final indexing feature lets you select elements by name. It assumes that the vector has a names attribute, defining a name for each element. This can be done by assigning a vector of character strings to the attribute:

```
> years <- c(1960, 1964, 1976, 1994)
> names(years) <- c("Kennedy", "Johnson", "Carter", "Clinton")
> years
Kennedy Johnson  Carter Clinton
   1960    1964    1976    1994
```

Once the names are defined, you can refer to individual elements by name:

```
> years["Carter"]
Carter
  1976
> years["Clinton"]
Clinton
   1994
```

This generalizes to allow indexing by vectors of names: R returns every element named in the index:

```
> years[c("Carter","Clinton")]
 Carter Clinton
   1976    1994
```

See Also

See Recipe 5.3 for more about the Recycling Rule.

2.10 Performing Vector Arithmetic

Problem

You want to operate on an entire vector at once.

Solution

The usual arithmetic operators can perform element-wise operations on entire vectors. Many functions operate on entire vectors, too, and return a vector result.

Discussion

Vector operations are one of R's great strengths. All the basic arithmetic operators can be applied to pairs of vectors. They operate in an element-wise manner; that is, the operator is applied to corresponding elements from both vectors:

```
> v <- c(11,12,13,14,15)
> w <- c(1,2,3,4,5)
> v + w
[1] 12 14 16 18 20
> v - w
[1] 10 10 10 10 10
> v * w
[1] 11 24 39 56 75
> v / w
[1] 11.000000  6.000000  4.333333  3.500000  3.000000
> w ^ v
[1]          1       4096    1594323  268435456 30517578125
```

Observe that the length of the result here is equal to the length of the original vectors. The reason is that each element comes from a pair of corresponding values in the input vectors.

If one operand is a vector and the other is a scalar, then the operation is performed between every vector element and the scalar:

```
> w
[1] 1 2 3 4 5
> w + 2
[1] 3 4 5 6 7
> w - 2
[1] -1  0  1  2  3
> w * 2
[1]  2  4  6  8 10
> w / 2
[1] 0.5 1.0 1.5 2.0 2.5
```

```
> w ^ 2
[1]  1  4  9 16 25
> 2 ^ w
[1]  2  4  8 16 32
```

For example, you can recenter an entire vector in one expression simply by subtracting the mean of its contents:

```
> w
[1] 1 2 3 4 5
> mean(w)
[1] 3
> w - mean(w)
[1] -2 -1  0  1  2
```

Likewise, you can calculate the z-score of a vector in one expression: subtract the mean and divide by the standard deviation:

```
> w
[1] 1 2 3 4 5
> sd(w)
[1] 1.581139
> (w - mean(w)) / sd(w)
[1] -1.2649111 -0.6324555  0.0000000  0.6324555  1.2649111
```

Yet the implementation of vector-level operations goes far beyond elementary arithmetic. It pervades the language, and many functions operate on entire vectors. The functions sqrt and log, for example, apply themselves to every element of a vector and return a vector of results:

```
> w
[1] 1 2 3 4 5
> sqrt(w)
[1] 1.000000 1.414214 1.732051 2.000000 2.236068
> log(w)
[1] 0.0000000 0.6931472 1.0986123 1.3862944 1.6094379
> sin(w)
[1]  0.8414710  0.9092974  0.1411200 -0.7568025 -0.9589243
```

There are two great advantages to vector operations. The first and most obvious is convenience. Operations that require looping in other languages are one-liners in R. The second is speed. Most vectorized operations are implemented directly in C code, so they are substantially faster than the equivalent R code you could write.

See Also

Performing an operation between a vector and a scalar is actually a special case of the Recycling Rule; see Recipe 5.3.

2.11 Getting Operator Precedence Right

Problem

Your R expression is producing a curious result, and you wonder if operator precedence is causing problems.

Solution

The full list of operators is shown in Table 2-1, listed in order of precedence from highest to lowest. Operators of equal precedence are evaluated from left to right except where indicated.

Table 2-1. Operator precedence

Operator	Meaning	See also
[[[Indexing	Recipe 2.9
:: :::	Access variables in a name space	
$ @	Component extraction, slot extraction	
^	Exponentiation (right to left)	
- +	Unary minus and plus	
:	Sequence creation	Recipe 2.7, Recipe 7.14
%any%	Special operators	Discussion
* /	Multiplication, division	Discussion
+ -	Addition, subtraction	
== != < > <= >=	Comparison	Recipe 2.8
!	Logical negation	
& &&	Logical "and", short-circuit "and"	
\| \|\|	Logical "or", short-circuit "or"	
~	Formula	Recipe 11.1
-> ->>	Rightward assignment	Recipe 2.2
=	Assignment (right to left)	Recipe 2.2
<- <<-	Assignment (right to left)	Recipe 2.2
?	Help	Recipe 1.7

Discussion

Getting your operator precedence wrong in R is a common problem. It certainly happens to me a lot. I unthinkingly expect that the expression `0:n-1` will create a sequence of integers from 0 to $n - 1$ but it does not:

```
> n <- 10
> 0:n-1
 [1] -1  0  1  2  3  4  5  6  7  8  9
```

It creates the sequence from –1 to *n* – 1 because R interprets it as (0:n)-1.

You might not recognize the notation %*any*% in the table. R interprets any text between percent signs (%...%) as a binary operator. Several such operators have predefined meanings:

%%
> Modulo operator

%/%
> Integer division

%*%
> Matrix multiplication

%in%
> Returns TRUE if the left operand occurs in its right operand; FALSE otherwise

You can also define new binary operators using the %...% notation; see Recipe 12.19. The point here is that all such operators have the same precedence.

See Also

See Recipe 2.10 for more about vector operations, Recipe 5.15 for more about matrix operations, and Recipe 12.19 to define your own operators. See the Arithmetic and Syntax topics in the R help pages as well as Chapters 5 and 6 of *R in a Nutshell* (O'Reilly).

2.12 Defining a Function

Problem

You want to define an R function.

Solution

Create a function by using the function keyword followed by a list of parameters and the function body. A one-liner looks like this:

```
function(param₁, ...., paramₙ) expr
```

The function body can be a series of expressions, in which case curly braces should be used around the function body:

```
function(param₁, ..., paramₙ) {
    expr₁
    .
    .
    .
    exprₘ
}
```

Discussion

Function definitions are how you tell R, "Here's how to calculate *this*." For example, R does not have a built-in function for calculating the coefficient of variation. You can define the calculation in this way:

```
> cv <- function(x) sd(x)/mean(x)
> cv(1:10)
[1] 0.5504819
```

The first line creates a function and assigns it to cv. The second line invokes the function, using 1:10 for the value of parameter x. The function returns the value of its single-expression body, sd(x)/mean(x).

After defining a function we can use it anywhere a function is expected, such as the second argument of lapply (Recipe 6.2):

```
> cv <- function(x) sd(x)/mean(x)
> lapply(lst, cv)
```

A multiline function uses curly braces to delimit the start and end of the function body. Here is a function that implements Euclid's algorithm for computing the greatest common divisor of two integers:

```
> gcd <- function(a,b) {
+     if (b == 0) return(a)
+     else return(gcd(b, a %% b))
+ }
```

R also allows *anonymous functions*; these are functions with no name that are useful for one-liners. The preceding example using cv and lapply can be shrunk to one line by using an anonymous function that is passed directly into lapply:

```
> lapply(lst, function(x) sd(x)/mean(x))
```

This is not a book about programming R, so I cannot fully cover the subtleties of coding an R function. But here are a few useful points:

Return value
> All functions return a value. Normally, a function returns the value of the last expression in its body. You can also use return(*expr*).

Call by value
> Function parameters are "call by value"—in other words, if you change a parameter then the change is local and does not affect the caller's value.

Local variables

You create a local variable simply by assigning a value to it. When the function exits, local variables are lost.

Conditional execution

The R syntax includes an `if` statement. See `help(Control)` for details.

Loops

The R syntax also includes `for` loops, `while` loops, and `repeat` loops. For details, see `help(Control)`.

Global variables

Within a function you can change a global variable by using the `<<-` assignment operator, but this is not encouraged.

See Also

For more about defining functions, see *An Introduction to R* (*http://cran.r-project.org/doc/manuals/R-intro.pdf*) and *R in a Nutshell*.

2.13 Typing Less and Accomplishing More

Problem

You are getting tired of typing long sequences of commands and especially tired of typing the same ones over and over.

Solution

Open an editor window and accumulate your reusable blocks of R commands there. Then, execute those blocks directly from that window. Reserve the command line for typing brief or one-off commands.

When you are done, you can save the accumulated code blocks in a script file for later use.

Discussion

The typical beginner to R types an expression and sees what happens. As he gets more comfortable, he types increasingly complicated expressions. Then he begins typing multiline expressions. Soon, he is typing the same multiline expressions over and over, perhaps with small variations, in order to perform his increasingly complicated calculations.

The experienced user does not often retype a complex expression. She may type the same expression once or twice, but when she realizes it is useful and reusable she will cut-and-paste it into an editor window of the GUI. To execute the snippet thereafter, she selects the snippet in the editor window and tells R to execute it, rather than retyping it. This technique is especially powerful as her snippets evolve into long blocks of code.

On Windows and OS X, a few features of the GUI facilitate this workstyle:

To open an editor window
From the main menu, select File → New script.

To execute one line of the editor window
Position the cursor on the line and then press Ctrl-R to execute it.

To execute several lines of the editor window
Highlight the lines using your mouse; then press Ctrl-R to execute them.

To execute the entire contents of the editor window
Press Ctrl-A to select the entire window contents and then Ctrl-R to execute them; or, from the main menu, select Edit → Run all.

Copying lines from the console window to the editor window is simply a matter of copy and paste. When you exit R, it will ask if you want to save the new script. You can either save it for future reuse or discard it.

I recently used this technique to create a small illustration of the Central Limit Theorem for some students. I wanted a probability density of the population overlaid with the density of the sample mean.

I opened R and then began defining variables and executing functions to create the plot of the population density. It dawned on me that I would likely reexecute those few commands, so I opened a script editor window in R and pasted the commands into it as shown in Figure 2-1.

I resumed typing commands in the console window, thinking I would overlay the existing graph with the density for the sample mean. However, the second density plot did not fit inside the graph. I needed to fix my function calls and reexecute everything.

Here came the big pay-off: I retyped little or nothing. Rather, I pasted a little more code into the editor window, tweaked the call to the plot function, and then reexecuted the whole window contents.

Voilà! The scaling problem was fixed. I tweaked the code again (to add titles and labels) and once again reexecuted the whole window contents; this yielded the final result shown in Figure 2-2. I saved the editor contents in a script file when I was done, enabling me to re-create this graph at will.

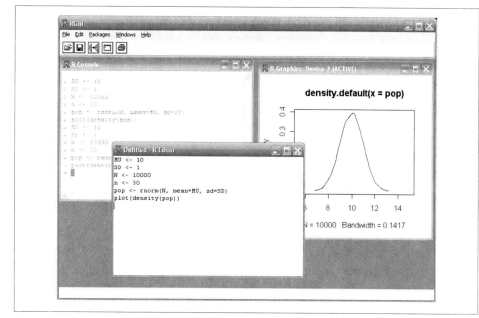

Figure 2-1. Capturing the commands in a new edit window

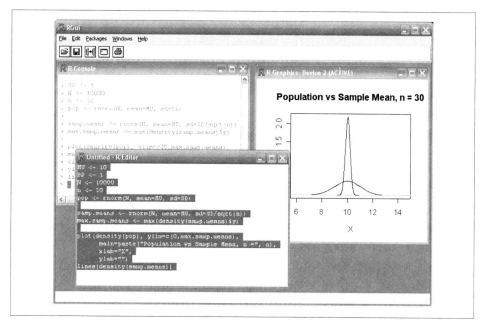

Figure 2-2. Evolved state of the code

2.14 Avoiding Some Common Mistakes

Problem

You want to avoid some of the common mistakes made by beginning users—and also by experienced users, for that matter.

Discussion

Here are some easy ways to make trouble for yourself:

Forgetting the parentheses after a function invocation

You call an R function by putting parentheses after the name. For instance, this line invokes the ls function:

```
> ls()
[1] "x" "y" "z"
```

However, if you omit the parentheses then R does not execute the function. Instead, it shows the function definition, which is almost never what you want:

```
> ls
function (name, pos = -1, envir = as.environment(pos), all.names = FALSE,
    pattern)
{
    if (!missing(name)) {
        nameValue <- try(name)
        if (identical(class(nameValue), "try-error")) {
            name <- substitute(name)
.
. (etc.)
.
```

Forgetting to double up backslashes in Windows file paths

This function call appears to read a Windows file called *F:\research\bio\assay.csv*, but it does not:

```
> tbl <- read.csv("F:\research\bio\assay.csv")
```

Backslashes (\) inside character strings have a special meaning and therefore need to be doubled up. R will interpret this file name as *F:researchbioassay.csv*, for example, which is not what the user wanted. See Recipe 4.5 for possible solutions.

Mistyping "<-" as "< (blank) -"

The assignment operator is <-, with no space between the < and the -:

```
> x <- pi      # Set x to 3.1415926...
```

If you accidentally insert a space between < and -, the meaning changes completely:

```
> x < - pi     # Oops! We are comparing x instead of setting it!
```

This is now a comparison (<) between *x* and negative π (- pi). It does not change x. If you are lucky, x is undefined and R will complain, alerting you that something is fishy:

```
> x < - pi
Error: object "x" not found
```

If x is defined, R will perform the comparison and print a logical value, TRUE or FALSE. That should alert you that something is wrong: an assignment does not normally print anything:

```
> x <- 0       # Initialize x to zero
> x < - pi     # Oops!
[1] FALSE
```

Incorrectly continuing an expression across lines

R reads your typing until you finish a complete expression, no matter how many lines of input that requires. It prompts you for additional input using the + prompt until it is satisfied. This example splits an expression across two lines:

```
> total <- 1 + 2 + 3 +     # Continued on the next line
+    4 + 5
> print(total)
[1] 15
```

Problems begin when you accidentally finish the expression prematurely, which can easily happen:

```
> total <- 1 + 2 + 3       # Oops! R sees a complete expression
>    + 4 + 5               # This is a new expression; R prints its value
[1] 9
> print(total)
[1] 6
```

There are two clues that something is amiss: R prompted you with a normal prompt (>), not the continuation prompt (+); and it printed the value of 4 + 5.

This common mistake is a headache for the casual user. It is a nightmare for programmers, however, because it can introduce hard-to-find bugs into R scripts.

Using = instead of ==

Use the double-equal operator (==) for comparisons. If you accidentally use the single-equal operator (=), you will irreversibly overwrite your variable:

```
> v == 0     # Compare v against zero
> v = 0      # Assign 0 to v, overwriting previous contents
```

Writing 1:n+1 *when you mean* 1:(n+1)

You might think that 1:n+1 is the sequence of numbers 1, 2, ..., *n*, *n* + 1. It's not. It is the sequence 1, 2, ..., *n* with 1 added to every element, giving 2, 3, ..., *n*, *n* + 1. This happens because R interprets 1:n+1 as (1:n)+1. Use parentheses to get exactly what you want:

```
> n <- 5
> 1:n+1
[1] 2 3 4 5 6
> 1:(n+1)
[1] 1 2 3 4 5 6
```

Getting bitten by the Recycling Rule

Vector arithmetic and vector comparisons work well when both vectors have the same length. However, the results can be baffling when the operands are vectors of differing lengths. Guard against this possibility by understanding and remembering the Recycling Rule, Recipe 5.3.

Installing a package but not loading it with `library()` *or* `require()`

Installing a package is the first step toward using it, but one more step is required. Use `library` or `require` to load the package into your search path. Until you do so, R will not recognize the functions or datasets in the package. See Recipe 3.6:

```
> truehist(x,n)
Error: could not find function "truehist"
> library(MASS)      # Load the MASS package into R
> truehist(x,n)
>
```

Writing `aList[i]` *when you mean* `aList[[i]]`, *or vice versa*

If the variable `lst` contains a list, it can be indexed in two ways: `lst[[n]]` is the *n*th element of the list; whereas `lst[n]` is a list whose only element is the *n*th element of `lst`. That's a big difference. See Recipe 5.7.

Using & *instead of* &&, *or vice versa; same for* | *and* ||

Use & and | in logical expressions involving the logical values TRUE and FALSE. See Recipe 2.9.

Use && and || for the flow-of-control expressions inside `if` and `while` statements.

Programmers accustomed to other programming languages may reflexively use && and || everywhere because "they are faster." But those operators give peculiar results when applied to vectors of logical values, so avoid them unless that's really what you want.

Passing multiple arguments to a single-argument function

What do you think is the value of `mean(9,10,11)`? No, it's not 10. It's 9. The `mean` function computes the mean of the first argument. The second and third arguments are being interpreted as other, positional arguments.

Some functions, such as `mean`, take one argument. Other arguments, such as `max` and `min`, take multiple arguments and apply themselves across all arguments. Be sure you know which is which.

Thinking that `max` *behaves like* `pmax`, *or that* `min` *behaves like* `pmin`

The `max` and `min` functions have multiple arguments and return one value: the maximum or minimum of all their arguments.

The pmax and pmin functions have multiple arguments but return a vector with values taken element-wise from the arguments. See Recipe 12.9.

Misusing a function that does not understand data frames

Some functions are quite clever regarding data frames. They apply themselves to the individual columns of the data frame, computing their result for each individual column. The mean and sd functions are good examples. These functions can compute the mean or standard deviation of each column because they understand that each column is a separate variable and that mixing their data is not sensible.

Sadly, not all functions are that clever. This includes the median, max, and min functions. They will lump together every value from every column and compute their result from the lump, which might not be what you want. Be aware of which functions are savvy to data frames and which are not.

Posting a question to the mailing list before searching for the answer

Don't waste your time. Don't waste other people's time. Before you post a question to a mailing list or to Stack Overflow, do your homework and search the archives. Odds are, someone has already answered your question. If so, you'll see the answer in the discussion thread for the question. See Recipe 1.12.

See Also

See Recipes 1.12, 2.9, 5.3, and 5.7.

Navigating the Software

Introduction

R is a big chunk of software, first and foremost. You will inevitably spend time doing what one does with any big piece of software: configuring it, customizing it, updating it, and fitting it into your computing environment. This chapter will help you perform those tasks. There is nothing here about numerics, statistics, or graphics. This is all about dealing with R as software.

3.1 Getting and Setting the Working Directory

Problem

You want to change your working directory. Or you just want to know what it is.

Solution

Command line

Use `getwd` to report the working directory, and use `setwd` to change it:

```
> getwd()
[1] "/home/paul/research"
> setwd("Bayes")
> getwd()
[1] "/home/paul/research/Bayes"
```

Windows

From the main menu, select File → Change dir... .

OS X

From the main menu, select Misc → Change Working Directory.

For both Windows and OS X, the menu selection opens the current working directory in a file browser. From there, you can navigate to a new working directory if desired.

Discussion

Your *working directory* is important because it is the default location for all file input and output—including reading and writing data files, opening and saving script files, and saving your workspace image. When you open a file and do not specify an absolute path, R will assume that the file is in your working directory.

The initial working directory depends upon how you started R. See Recipe 1.2.

See Also

See Recipe 4.5 for dealing with filenames in Windows.

3.2 Saving Your Workspace

Problem

You want to save your workspace without exiting from R.

Solution

Call the `save.image` function:

```
> save.image()
```

Discussion

Your workspace holds your R variables and functions, and it is created when R starts. The workspace is held in your computer's main memory and lasts until you exit from R, at which time you can save it.

However, you may want to save your workspace without exiting R. You might go to lunch, for example, and want to protect your work against an unexpected power outage or machine crash. Use the `save.image` function.

The workspace is written to a file called *.RData* in the working directory. When R starts, it looks for that file and, if found, initializes the workspace from it.

A sad fact is that the workspace does not include your open graphs: that cool graph on your screen disappears when you exit R, and there is no simple way to save and restore it. So before you exit, save the data and the R code that will re-create your graphs.

See Also

See Recipe 1.2 for how to save your workspace when exiting R and Recipe 3.1 for setting the working directory.

3.3 Viewing Your Command History

Problem

You want to see your recent sequence of commands.

Solution

Scroll backward by pressing the up arrow or Ctrl-P. Or use the `history` function to view your most recent input:

```
> history()
```

Discussion

The `history` function will display your most recent commands. By default it shows the most recent 25 lines, but you can request more:

```
> history(100)          # Show 100 most recent lines of history
> history(Inf)          # Show entire saved history
```

For very recent commands, simply scroll backward through your input using the command-line editing feature: pressing either the up arrow or Ctrl-P will cause your previous typing to reappear, one line at a time.

If you've exited from R then you can still see your command history. It saves the history in a file called *.Rhistory* in the working directory, if requested. Open the file with a text editor and then scroll to the bottom; you will see your most recent typing.

3.4 Saving the Result of the Previous Command

Problem

You typed an expression into R that calculated the value, but you forgot to save the result in a variable.

Solution

A special variable called `.Last.value` saves the value of the most recently evaluated expression. Save it to a variable before you type anything else.

Discussion

It is frustrating to type a long expression or call a long-running function but then forget to save the result. Fortunately, you needn't retype the expression nor invoke the function again—the result was saved in the `.Last.value` variable:

```
> aVeryLongRunningFunction()        # Oops! Forgot to save the result!
[1] 147.6549
> x <- .Last.value                  # Capture the result now
> x
[1] 147.6549
```

A word of caution: the contents of `.Last.value` are overwritten every time you type another expression, so capture the value immediately. If you don't remember until another expression has been evaluated, it's too late.

See Also

See Recipe 3.3 to recall your command history.

3.5 Displaying the Search Path

Problem

You want to see the list of packages currently loaded into R.

Solution

Use the search function with no arguments:

```
> search()
```

Discussion

The *search path* is a list of packages that are currently loaded into memory and available for use. Although many packages may be installed on your computer, only a few of them are actually loaded into the R interpreter at any given moment. You might be wondering which packages are loaded right now.

With no arguments, the search function returns the list of loaded packages. It produces an output like this:

```
> search()
[1] ".GlobalEnv"          "package:stats"      "package:graphics"
[4] "package:grDevices"   "package:utils"      "package:datasets"
[7] "package:methods"     "Autoloads"          "package:base"
```

Your machine may return a different result, depending on what's installed there. The return value of search is a vector of strings. The first string is ".GlobalEnv", which refers to your workspace. Most strings have the form "package:packagename", which indicates that the package called *packagename* is currently loaded into R. In this example, the loaded packages include stats, graphics, grDevices, utils, and so forth.

R uses the search path to find functions. When you type a function name, R searches the path—in the order shown—until it finds the function in a loaded package. If the function is found, R executes it. Otherwise, it prints an error message and stops. (There

is actually a bit more to it: the search path can contain environments, not just packages, and the search algorithm is different when initiated by an object within a package; see the *R Language Definition* for details.)

Since your workspace (.GlobalEnv) is first in the list, R looks for functions in your workspace before searching any packages. If your workspace and a package both contain a function with the same name, your workspace will "mask" the function; this means that R stops searching after it finds your function and so never sees the package function. This is a blessing if you want to override the package function...and a curse if you still want access to the package function.

R also uses the search path to find R datasets (not files) via a similar procedure.

Unix users: don't confuse the R search path with the Unix search path (the PATH environment variable). They are conceptually similar but two distinct things. The R search path is internal to R and is used by R only to locate functions and datasets, whereas the Unix search path is used by Unix to locate executable programs.

See Also

See Recipe 3.6 for loading packages into R, Recipe 3.8 for the list of installed packages (not just loaded packages), and Recipe 5.31 for inserting data frames into the search path.

3.6 Accessing the Functions in a Package

Problem

A package installed on your computer is either a standard package or a package downloaded by you. When you try using functions in the package, however, R cannot find them.

Solution

Use either the library function or the require function to load the package into R:

```
> library(packagename)
```

Discussion

R comes with several standard packages, but not all of them are automatically loaded when you start R. Likewise, you can download and install many useful packages from CRAN, but they are not automatically loaded when you run R. The MASS package comes standard with R, for example, but you could get this message when using the lda function in that package:

```
> lda(x)
Error: could not find function "lda"
```

R is complaining that it cannot find the lda function among the packages currently loaded into memory.

When you use the library function or the require function, R loads the package into memory and its contents become immediately available to you:

```
> lda(f ~ x + y)
Error: could not find function "lda"
> library(MASS)                        # Load the MASS library into memory
> lda(f ~ x + y)                       # Now R can find the function
Call:
lda(f ~ x + y)

Prior probabilities of groups:
.
. (etc.)
.
```

Before calling library, R does not recognize the function name. Afterward, the package contents are available and calling the lda function works.

Notice that you needn't enclose the package name in quotes.

The require function is nearly identical to library. It has two features that are useful for writing scripts. It returns TRUE if the package was successfully loaded and FALSE otherwise. It also generates a mere warning if the load fails—unlike library, which generates an error.

Both functions have a key feature: they do not reload packages that are already loaded, so calling twice for the same package is harmless. This is especially nice for writing scripts. The script can load needed packages while knowing that loaded packages will not be reloaded.

The detach function will unload a package that is currently loaded:

```
> detach(package:MASS)
```

Observe that the package name must be qualified, as in package:MASS.

One reason to unload a package is that it contains a function whose name conflicts with a same-named function lower on the search list. When such a conflict occurs, we say the higher function *masks* the lower function. You no longer "see" the lower function because R stops searching when it finds the higher function. Hence unloading the higher package unmasks the lower name.

See Also

See Recipe 3.5.

3.7 Accessing Built-in Datasets

Problem

You want to use one of R's built-in datasets.

Solution

The standard datasets distributed with R are already available to you, since the datasets package is in your search path.

To access datasets in other packages, use the data function while giving the dataset name and package name:

```
> data(dsname, package="pkgname")
```

Discussion

R comes with many built-in datasets. These datasets are useful when you are learning about R, since they provide data with which to experiment.

Many datasets are kept in a package called (naturally enough) datasets, which is distributed with R. That package is in your search path, so you have instant access to its contents. For example, you can use the built-in dataset called pressure:

```
> head(pressure)
  temperature pressure
1           0   0.0002
2          20   0.0012
3          40   0.0060
4          60   0.0300
5          80   0.0900
6         100   0.2700
```

If you want to know more about pressure, use the help function to learn about it and other datasets:

```
> help(pressure)        # Bring up help page for pressure dataset
```

You can see a table of contents for datasets by calling the data function with no arguments:

```
> data()                # Bring up a list of datasets
```

Any R package can elect to include datasets that supplement those supplied in datasets. The MASS package, for example, includes many interesting datasets. Use the data function to access a dataset in a specific package by using the package argument. MASS includes a dataset called Cars93, which you can access in this way:

```
> data(Cars93, package="MASS")
```

After this call to data, the Cars93 dataset is available to you; then you can execute summary(Cars93), head(Cars93), and so forth.

When attaching a package to your search list (e.g., via library(MASS)), you don't need to call data. Its datasets become available automatically when you attach it.

You can see a list of available datasets in MASS, or any other package, by using the data function with a package argument and no dataset name:

```
> data(package="pkgname")
```

See Also

See Recipe 3.5 for more about the search path and Recipe 3.6 for more about packages and the library function.

3.8 Viewing the List of Installed Packages

Problem

You want to know what packages are installed on your machine.

Solution

Use the library function with no arguments for a basic list. Use installed.packages to see more detailed information about the packages.

Discussion

The library function with no arguments prints a list of installed packages. The list can be quite long. On a Linux computer, these might be the first few lines of output:

```
> library()
Packages in library '/usr/local/lib/R/site-library':

boot                    Bootstrap R (S-Plus) Functions (Canty)
CGIwithR                CGI Programming in R
class                   Functions for Classification
cluster                 Cluster Analysis Extended Rousseeuw et al.
DBI                     R Database Interface
expsmooth               Data sets for "Forecasting with exponential
                        smoothing"
.
. (etc.)
.
```

On Windows and OS X, the list is displayed in a pop-up window.

You can get more details via the installed.packages function, which returns a matrix of information regarding the packages on your machine. Each matrix row corresponds to one installed package. The columns contain the information such as package name, library path, and version. The information is taken from R's internal database of installed packages.

To extract useful information from this matrix, use normal indexing methods. This Windows snippet calls `installed.packages` and extracts both the `Package` and `Version` columns, letting you see what version of each package is installed:

```
> installed.packages()[,c("Package","Version")]
          Package       Version
acepack   "acepack"     "1.3-2.2"
alr3      "alr3"        "1.0.9"
base      "base"        "2.4.1"
boot      "boot"        "1.2-27"
bootstrap "bootstrap"   "1.0-20"
calibrate "calibrate"   "0.0"
car       "car"         "1.2-1"
chron     "chron"       "2.3-12"
class     "class"       "7.2-30"
cluster   "cluster"     "1.11.4"
.
. (etc.)
.
```

See Also

See Recipe 3.6 for loading a package into memory.

3.9 Installing Packages from CRAN

Problem

You found a package on CRAN, and now you want to install it on your computer.

Solution

Command line

Use the `install.packages` function, putting the name of the package in quotes:

```
> install.packages("packagename")
```

Windows

You can also download and install via Packages → Install package(s)... from the main menu.

OS X

You can also download and install via Packages & Data → Package Installer from the main menu.

On all platforms, you will be asked to select a CRAN mirror.

On Windows and OS X, you will also be asked to select the packages for download.

System-wide Installation on Linux or Unix

On Linux or Unix systems, root privileges are required to install packages into the system-wide libraries because those directories are not usually writable by mere mortals. For that reason, installations are usually performed by a system administrator. If the administrator is unwilling or unable to do a system-wide installation, you can still install packages in your personal library.

If you have root privileges:

1. Run su or sudo to start a root shell.
2. Start an R session in the root shell.
3. From there, execute the install.packages function.

If you don't have root privileges, you'll know very quickly. The install.packages function will stop, warning you that it cannot write into the necessary directories. Then it will ask if you want to create a personal library instead. If you do, it will create the necessary directories in your home directory and install the package there.

Discussion

Installing a package locally is the first step toward using it. The installer will prompt you for a mirror site from which it can download the package files:

```
--- Please select a CRAN mirror for use in this session ---
```

It will then display a list of CRAN mirror sites. Select one close to you.

The official CRAN server is a relatively modest machine generously hosted by the Department of Statistics and Mathematics at WU Wien, Vienna, Austria. If every R user downloaded from the official server, it would buckle under the load, so there are numerous mirror sites around the globe. You are strongly encouraged to find and use a nearby mirror.

If the new package depends upon other packages that are not already installed locally, then the R installer will automatically download and install those required packages. This is a huge benefit that frees you from the tedious task of identifying and resolving those dependencies.

There is a special consideration when installing on Linux or Unix. You can install the package either in the system-wide library or in your personal library. Packages in the system-wide library are available to everyone; packages in your personal library are (normally) used only by you. So a popular, well-tested package would likely go in the system-wide library whereas an obscure or untested package would go into your personal library.

By default, install.packages assumes you are performing a system-wide install. To install into your personal library, first create a directory for the library—for example, ~/lib/R:

```
$ mkdir ~/lib
$ mkdir ~/lib/R
```

Set the R_LIBS environment variable before starting R. Otherwise, it will not know your library's location:

```
$ export R_LIBS=~/lib/R        # bash syntax
$ setenv R_LIBS ~/lib/R        # csh syntax
```

Then, call install.packages with the lib argument set to that directory:

```
> install.packages("packagename", lib="~/lib/R")
```

See Also

See Recipe 1.11 for ways to find relevant packages and Recipe 3.6 for using a package after installing it.

3.10 Setting a Default CRAN Mirror

Problem

You are downloading packages. You want to set a default CRAN mirror so R does not prompt for one each time.

Solution

This solution assumes you have an *.Rprofile*, as described in Recipe 3.16:

1. Call the chooseCRANmirror function:

   ```
   > chooseCRANmirror()
   ```

 R will present a list of CRAN mirrors.

2. Select a CRAN mirror from the list and press OK.

3. To get the URL of the mirror, look at the first element of the repos option:

   ```
   > options("repos")[[1]][1]
   ```

4. Add this line to your *.Rprofile* file:

   ```
   options(repos="URL")
   ```

 where URL is the URL of the mirror.

Discussion

When you install packages, you probably use the same CRAN mirror each time (namely, the mirror closest to you). You probably get tired of R repeatedly asking you to select a mirror. If you follow the Solution, R will stop prompting for a mirror because it will have a default mirror.

The repos option is the name of your default mirror. The chooseCRANmirror function has the important side effect of setting the repos option according to your selection. The problem is that R forgets the setting when it exits, leaving no permanent default. By setting repos in your .Rprofile, you restore the setting every time R starts.

See Also

See Recipe 3.16 for more about the .Rprofile file and the options function.

3.11 Suppressing the Startup Message

Problem

You are tired of seeing R's verbose startup message.

Solution

Use the --quiet command-line option when you start R.

Discussion

The startup message from R is handy for beginners because it contains useful information about the R project and getting help. But the novelty wears off pretty quickly.

If you start R from the shell prompt, use the --quiet option to hide the startup message:

```
$ R --quiet
>
```

On my Linux box, I aliased R like this so I never see the startup message:

```
$ alias R="/usr/bin/R --quiet"
```

If you are starting R in Windows using a shortcut, you can embed the --quiet option inside the shortcut. Right-click on the shortcut icon; select Properties; select the Shortcut tab; and, at the end of the Target string, add the --quiet option. Be careful to leave a space between the end of the program path and --quiet.

See Also

See Recipe 1.2.

3.12 Running a Script

Problem

You captured a series of R commands in a text file. Now you want to execute them.

Solution

The source function instructs R to read the text file and execute its contents:

```
> source("myScript.R")
```

Discussion

When you have a long or frequently used piece of R code, capture it inside a text file. That lets you easily rerun the code without having to retype it. Use the source function to read and execute the code, just as if you had typed it into the R console.

Suppose the file *hello.R* contains this one, familiar greeting:

```
print("Hello, World!")
```

Then sourcing the file will execute the file contents:

```
> source("hello.R")
[1] "Hello, World!"
```

Setting echo=TRUE will echo the script lines before they are executed, with the R prompt shown before each line:

```
> source("hello.R", echo=TRUE)

> print("Hello, World!")
[1] "Hello, World!"
```

See Also

See Recipe 2.13 for running blocks of R code inside the GUI.

3.13 Running a Batch Script

Problem

You are writing a command script, such as a shell script in Unix or OS X or a BAT script in Windows. Inside your script, you want to execute an R script.

Solution

Run the R program with the CMD BATCH subcommand, giving the script name and the output file name:

```
$ R CMD BATCH scriptfile outputfile
```

If you want the output sent to stdout or if you need to pass command-line arguments to the script, consider the Rscript command instead:

```
$ Rscript scriptfile arg1 arg2 arg3
```

Discussion

R is normally an interactive program, one that prompts the user for input and then displays the results. Sometimes you want to run R in batch mode, reading commands from a script. This is especially useful inside shell scripts, such as scripts that include a statistical analysis.

The `CMD BATCH` subcommand puts R into batch mode, reading from *scriptfile* and writing to *outputfile*. It does not interact with a user.

You will likely use command-line options to adjust R's batch behavior to your circumstances. For example, using `--quiet` silences the startup messages that would otherwise clutter the output:

```
$ R CMD BATCH --quiet myScript.R results.out
```

Other useful options in batch mode include the following:

`--slave`
> Like `--quiet`, but it makes R even more silent by inhibiting echo of the input.

`--no-restore`
> At startup, do not restore the R workspace. This is important if your script expects R to begin with an empty workspace.

`--no-save`
> At exit, do not save the R workspace. Otherwise, R will save its workspace and overwrite the *.RData* file in the working directory.

`--no-init-file`
> Do not read either the *.Rprofile* or *~/.Rprofile* files.

The `CMD BATCH` subcommand normally calls `proc.time` when your script completes, showing the execution time. If this annoys you then end your script by calling the q function with `runLast=FALSE`, which will prevent the call to `proc.time`.

The `CMD BATCH` subcommand has two limitations: the output always goes to a file, and you cannot easily pass command-line arguments to your script. If either limitation is a problem, consider using the `Rscript` program that comes with R. The first command-line argument is the script name, and the remaining arguments are given to the script:

```
$ Rscript myScript.R arg1 arg2 arg3
```

Inside the script, the command-line arguments can be accessed by calling `commandArgs`, which returns the arguments as a vector of strings:

```
argv <- commandArgs(TRUE)
```

The `Rscript` program takes the same command-line options as `CMD BATCH`, which were just described.

Output is written to stdout, which R inherits from the calling shell script, of course. You can redirect the output to a file by using the normal redirection:

```
$ Rscript --slave myScript.R arg1 arg2 arg3 >results.out
```

Here is a small R script, *arith.R*, that takes two command-line arguments and performs four arithmetic operations on them:

```
argv <- commandArgs(TRUE)
x <- as.numeric(argv[1])
y <- as.numeric(argv[2])

cat("x =", x, "\n")
cat("y =", y, "\n")
cat("x + y = ", x + y, "\n")
cat("x - y = ", x - y, "\n")
cat("x * y = ", x * y, "\n")
cat("x / y = ", x / y, "\n")
```

The script is invoked like this:

```
$ Rscript arith.R 2 3.1415
```

which produces the following output:

```
x = 2
y = 3.1415
x + y = 5.1415
x - y = -1.1415
x * y = 6.283
x / y = 0.6366385
```

On Linux or Unix, you can make the script fully self-contained by placing a #! line at the head with the path to the Rscript program. Suppose that Rscript is installed in */usr/bin/Rscript* on your system. Then adding this line to *arith.R* makes it a self-contained script:

```
#!/usr/bin/Rscript --slave

argv <- commandArgs(TRUE)
x <- as.numeric(argv[1])
.
. (etc.)
.
```

At the shell prompt, we mark the script as executable:

```
$ chmod +x arith.R
```

Now we can invoke the script directly without the Rscript prefix:

```
$ arith.R 2 3.1415
```

See Also

See Recipe 3.12 for running a script from within R.

3.14 Getting and Setting Environment Variables

Problem

You want to see the value of an environment variable, or you want to change its value.

Solution

Use the `Sys.getenv` function to see values. Use `Sys.putenv` to change them:

```
> Sys.getenv("SHELL")
      SHELL
"/bin/bash"
> Sys.setenv(SHELL="/bin/ksh")
```

Discussion

Environment variables are often used on Linux and Unix to configure and control the software. Each process has its own set of environment variables, which are inherited from its parent process. You sometimes need to see the environment variable settings for your R process in order to understand its behavior. Likewise, you sometimes need to change those settings to modify that behavior.

Sometimes I start R in one place but actually want the graphics to appear in a different place. For example, I might run R on one Linux terminal but want the graphics to appear on a larger display that an audience can see more easily. Or I might want to run R on my Linux workstation and have the graphics appear on a colleague's workstation. On Linux, R uses the X Windows system for its graphics displays. X Windows chooses its display device according to an environment variable called `DISPLAY`. We can use `Sys.getenv` to see its value:

```
> Sys.getenv("DISPLAY")
DISPLAY
 ":0.0"
```

All environment variables are string-valued, and here the value is ":0.0"—an obscure way of saying that the R session is connected to display 0, screen number 0 on my workstation.

To redirect graphics to the local display at 10.0, we can change `DISPLAY` in this way:

```
> Sys.putenv(DISPLAY="localhost:10.0")
```

Similarly, we could redirect graphics to display 0, screen 0 on the workstation called zeus:

```
> Sys.putenv(DISPLAY="zeus:0.0")
```

In either event, you must change the `DISPLAY` variable before drawing your graphic.

3.15 Locating the R Home Directory

Problem

You need to know the R home directory, which is where the configuration and installation files are kept.

Solution

R creates an environment variable called R_HOME that you can access by using the Sys.getenv function:

```
> Sys.getenv("R_HOME")
```

Discussion

Most users will never need to know the R home directory. But system administrators or sophisticated users must know in order to check or change the R installation files.

When R starts, it defines an *environment* variable (not an R variable) called R_HOME, which is the path to the R home directory. The Sys.getenv function can retrieve its value. Here are examples by platform. The exact value reported will almost certainly be different on your own computer:

On Windows

```
> Sys.getenv("R_HOME")
                      R_HOME
"C:\\PROGRA~1\\R\\R-21~1.1"
```

On OS X

```
> Sys.getenv("R_HOME")
"/Library/Frameworks/R.framework/Resources"
```

On Linux or Unix

```
> Sys.getenv("R_HOME")
     R_HOME
"/usr/lib/R"
```

The Windows result looks funky because R reports the old, DOS-style compressed path name. The full, user-friendly path would be *C:\Program Files\R\R-2.10.1* in this case.

On Unix and OS X, you can also run the R program from the shell and use the RHOME subcommand to display the home directory:

```
$ R RHOME
/usr/lib/R
```

Note that the R home directory on Unix and OS X contains the installation files but not necessarily the R executable file. The executable could be in */usr/bin* while the R home directory is, for example, */usr/lib/R*.

3.16 Customizing R

Problem

You want to customize your R sessions by, for instance, changing configuration options or preloading packages.

Solution

Create a script called *.Rprofile* that customizes your R session. R will execute the *.Rprofile* script when it starts. The placement of *.Rprofile* depends upon your platform:

OS X, Linux, or Unix
 Save the file in your home directory (*~/.Rprofile*).

Windows
 Save the file in your *My Documents* directory.

Discussion

R executes *profile scripts* when it starts, freeing you from repeatedly loading often-used packages or tweaking the R configuration options.

You can create a profile script called *.Rprofile* and place it in your home directory (OS X, Linux, Unix) or your *My Documents* directory (Windows XP) or *Documents* directory (Windows Vista, Windows 7). The script can call functions to customize your sessions, such as this simple script that loads the MASS package and sets the prompt to R>:

```
require(MASS)
options(prompt="R> ")
```

The profile script executes in a bare-bones environment, so there are limits on what it can do. Trying to open a graphics window will fail, for example, because the graphics package is not yet loaded. Also, you should not attempt long-running computations.

You can customize a particular project by putting an *.Rprofile* file in the directory that contains the project files. When R starts in that directory, it reads the local *.Rprofile* file; this allows you to do project-specific customizations (e.g., loading packages needed only by the project). However, if R finds a local profile then it does *not* read the global profile. That can be annoying, but it's easily fixed: simply source the global profile from the local profile. On Unix, for instance, this local profile would execute the global profile first and then execute its local material:

```
source("~/.Rprofile")
#
# ... remainder of local .Rprofile...
#
```

Setting Options

Some customizations are handled via calls to the options function, which sets the R configuration options. There are many such options, and the R help page for options lists them all:

```
> help(options)
```

Here are some examples:

browser="*path*"
 Path of default HTML browser

digits=*n*
 Suggested number of digits to print when printing numeric values

editor="*path*"
 Default text editor

prompt="*string*"
 Input prompt

repos="*url*"
 URL for default repository for packages

warn=*n*
 Controls display of warning messages

Loading Packages

Another common customization is the preloading of packages. You may want certain packages loaded every time you run R because you use them so often. A simple way to accomplish this is by calling the **require** function inside your .*Rprofile* script, like this call (which loads the **tseries** package):

```
require(tseries)
```

If **require** generates warning messages that are polluting your output when R starts, surround it with a **suppressMessages** function. That'll muzzle it:

```
suppressMessages(require(tseries))
```

Instead of calling **require** explicitly, you can set a configuration parameter called **defaultPackages**. This is the list of packages loaded by R at startup. It initially contains a system-defined list. If you append package names to the list then R will load them, too, saving you from having to call **library** or **require** yourself.

Here is a clip from my R profile that adjusts the list of loaded packages. I almost always use the **zoo** package, so I append it to the **defaultPackages** list:

```
pkgs <- getOption("defaultPackages")    # Get system list of packages to load
pkgs <- c(pkgs, "zoo")                  # Append "zoo" to list
options(defaultPackages = pkgs)         # Update configuration option
rm(pkgs)                                # Clean up our temp variable
```

One advantage of using defaultPackages is that you can control the placement of the packages in the search list. Packages appended to the end are loaded last, so in this example the zoo package will appear at the head of the search list. (Remember that packages are inserted at the head of the search list when they are loaded.)

The price for always loading the zoo package is that R starts a little more slowly, even when I don't really need zoo. In my case, I just was fed up with typing library(zoo) nearly every time I ran R, so I decided the slower startup was a good trade-off for the convenience.

Startup Sequence

Here is a simplified overview of what happens when R starts (type help(Startup) to see the full details):

1. *R executes the Rprofile.site script.* This is the site-level script that enables system administrators to override default options with localizations. The script's full path is R_HOME/*etc/Rprofile.site*. (R_HOME is the R home directory; see Recipe 3.15.)

 The R distribution does not include an *Rprofile.site* file. Rather, the system administrator creates one if it is needed.

2. *R executes the .Rprofile script in the working directory; or, if that file does not exist, executes the .Rprofile script in your home directory.* This is the user's opportunity to customize R for his or her purposes. The .Rprofile script in the home directory is used for global customizations. The .Rprofile script in a lower-level directory can perform specific customizations when R is started there; for instance, customizing R when started in a project-specific directory.

3. *R loads the workspace saved in .RData, if that file exists in the working directory.* R saves your workspace in the file called *.RData* when it exits. It reloads your workspace from that file, restoring access to your local variables and functions.

4. *R executes the .First function, if you defined one.* The .First function is a useful place for users or projects to define startup initialization code. You can define it in your *.Rprofile* or in your workspace.

5. *R executes the .First.sys function.* This step loads the default packages. The function is internal to R and not normally changed by either users or administrators.

Observe that R does not load the default packages until the final step, when it executes the .First.sys function. Before that, only the base package has been loaded. This is a key fact because it means the previous steps cannot assume that packages other than the base are available. It also explains why trying to open a graphical window in your *.Rprofile* script fails: the graphics packages aren't loaded yet.

See Also

See Recipe 3.6 for more about loading packages. See the R help page for Startup (help(Startup)) and the R help page for options (help(options)).

Input and Output

Introduction

All statistical work begins with data, and most data is stuck inside files and databases. Dealing with input is probably the first step of implementing any significant statistical project.

All statistical work ends with reporting numbers back to a client, even if you are the client. Formatting and producing output is probably the climax of your project.

Casual R users can solve their input problems by using basic functions such as `read.csv` to read CSV files and `read.table` to read more complicated, tabular data. They can use `print`, `cat`, and `format` to produce simple reports.

Users with heavy-duty input/output (I/O) needs are strongly encouraged to read the *R Data Import/Export* guide, available on CRAN at *http://cran.r-project.org/doc/manuals/R-data.pdf*. This manual includes important information on reading data from sources such as spreadsheets, binary files, other statistical systems, and relational databases.

A Philosophical Note

Several of my Statistical Analysis System (SAS) friends are disappointed with the input facilities of R. They point out that SAS has an elaborate set of commands for reading and parsing input files in many formats. R does not, and this leads them to conclude that R is not ready for real work. After all, if it can't read your data, what good is it?

I think they do not understand the design philosophy behind R, which is based on a statistical package called S. The authors of S worked at Bell Labs and were steeped in the Unix design philosophy. A keystone of that philosophy is the idea of *modular tools*. Programs in Unix are not large, monolithic programs that try to do everything. Instead, they are smaller, specialized tools that each do one thing well. The Unix user joins the programs together like building blocks, creating systems from the components.

R does statistics and graphics well. Very well, in fact. It is superior in that way to many commercial packages.

R is not a great tool for preprocessing data files, however. The authors of S assumed you would perform that munging with some other tool: perl, awk, sed, cut, paste, whatever floats your boat. Why should they duplicate that capability?

If your data is difficult to access or difficult to parse, consider using an outboard tool to preprocess the data before loading it into R. Let R do what R does best.

4.1 Entering Data from the Keyboard

Problem

You have a small amount of data, too small to justify the overhead of creating an input file. You just want to enter the data directly into your workspace.

Solution

For very small datasets, enter the data as literals using the c() constructor for vectors:

```
> scores <- c(61, 66, 90, 88, 100)
```

Alternatively, you can create an empty data frame and then invoke the built-in, spreadsheet-like editor to populate it:

```
> scores <- data.frame()     # Create empty data frame
> scores <- edit(score)      # Invoke editor, overwrite with edited data
```

On Windows, you can also invoke the editor by selecting Edit → Data frame… from the menu.

Discussion

When I am working on a toy problem, I don't want the hassle of creating and then reading the data file. I just want to enter the data into R. The easiest way is by using the c() constructor for vectors, as shown in the Solution.

This approach works for data frames, too, by entering each variable (column) as a vector:

```
> points <- data.frame(
+              label=c("Low", "Mid", "High"),
+              lbound=c(    0, 0.67,   1.64),
+              ubound=c(0.674, 1.64,   2.33)
+          )
```

See Also

See Recipe 5.26 for more about using the built-in data editor, as suggested in the Solution.

4.2 Printing Fewer Digits (or More Digits)

Problem

Your output contains too many digits or too few digits. You want to print fewer or more.

Solution

For `print`, the `digits` parameter can control the number of printed digits.

For `cat`, use the `format` function (which also has a `digits` parameter) to alter the formatting of numbers.

Discussion

R normally formats floating-point output to have seven digits:

```
> pi
[1] 3.141593
> 100*pi
[1] 314.1593
```

This works well most of the time but can become annoying when you have lots of numbers to print in a small space. It gets downright misleading when there are only a few significant digits in your numbers and R still prints seven.

The `print` function lets you vary the number of printed digits using the `digits` parameter:

```
> print(pi, digits=4)
[1] 3.142
> print(100*pi, digits=4)
[1] 314.2
```

The cat function does not give you direct control over formatting. Instead, use the `format` function to format your numbers before calling `cat`:

```
> cat(pi, "\n")
3.141593
> cat(format(pi,digits=4), "\n")
3.142
```

This is R, so both `print` and `format` will format entire vectors at once:

```
> pnorm(-3:3)
[1] 0.001349898 0.022750132 0.158655254 0.500000000 0.841344746 0.977249868
[7] 0.998650102
> print(pnorm(-3:3), digits=3)
[1] 0.00135 0.02275 0.15866 0.50000 0.84134 0.97725 0.9986
```

Notice that `print` formats the vector elements consistently: finding the number of digits necessary to format the smallest number and then formatting all numbers to have the same width (though not necessarily the same number of digits). This is extremely useful for formating an entire table:

```
> q <- seq(from=0,to=3,by=0.5)
> tbl <- data.frame(Quant=q, Lower=pnorm(-q), Upper=pnorm(q))
> tbl                          # Unformatted print
  Quant      Lower      Upper
1   0.0 0.500000000 0.5000000
2   0.5 0.308537539 0.6914625
3   1.0 0.158655254 0.8413447
4   1.5 0.066807201 0.9331928
5   2.0 0.022750132 0.9772499
6   2.5 0.006209665 0.9937903
7   3.0 0.001349898 0.9986501
> print(tbl,digits=2)          # Formatted print: fewer digits
  Quant Lower Upper
1   0.0 0.5000  0.50
2   0.5 0.3085  0.69
3   1.0 0.1587  0.84
4   1.5 0.0668  0.93
5   2.0 0.0228  0.98
6   2.5 0.0062  0.99
7   3.0 0.0013  1.00
```

You can also alter the format of *all* output by using the options function to change the default for digits:

```
> pi
[1] 3.141593
> options(digits=15)
> pi
[1] 3.14159265358979
```

But this is a poor choice in my experience, since it also alters the output from R's built-in functions, and that alteration will likely be unpleasant.

See Also

Other functions for formatting numbers include sprintf and formatC; see their help pages for details.

4.3 Redirecting Output to a File

Problem

You want to redirect the output from R into a file instead of your console.

Solution

You can redirect the output of the cat function by using its file argument:

```
> cat("The answer is", answer, "\n", file="filename")
```

Use the sink function to redirect *all* output from both print and cat. Call sink with a filename argument to begin redirecting console output to that file. When you are done, use sink with no argument to close the file and resume output to the console:

```
> sink("filename")                    # Begin writing output to file

. . . other session work . . .

> sink()                              # Resume writing output to console
```

Discussion

The print and cat functions normally write their output to your console. The cat function writes to a file if you supply a file argument, which can be either a filename or a connection. The print function cannot redirect its output, but the sink function can force all output to a file. A common use for sink is to capture the output of an R script:

```
> sink("script_output.txt")          # Redirect output to file
> source("script.R")                 # Run the script, capturing its output
> sink()                             # Resume writing output to console
```

If you are repeatedly cating items to one file, be sure to set append=TRUE. Otherwise, each call to cat will simply overwrite the file's contents:

```
cat(data, file="analysisReport.out")
cat(results, file="analysisRepart.out", append=TRUE)
cat(conclusion, file="analysisReport.out", append=TRUE)
```

Hard-coding file names like this is a tedious and error-prone process. Did you notice that the filename is misspelled in the second line? Instead of hard-coding the filename repeatedly, I suggest opening a connection to the file and writing your output to the connection:

```
con <- file("analysisReport.out", "w")
cat(data, file=con)
cat(results, file=con)
cat(conclusion, file=con)
close(con)
```

(You don't need append=TRUE when writing to a connection because it's implied.) This technique is especially valuable inside R scripts because it makes your code more reliable and more maintainable.

4.4 Listing Files

Problem

You want to see a listing of your files without the hassle of switching to your file browser.

Solution

The list.files function shows the contents of your working directory:

```
> list.files()
```

Discussion

This function is just plain handy. If I can't remember the name of my data file (was it *sample_data.csv* or *sample-data.csv*?), I do a quick list.files to refresh my memory:

```
> list.files()
[1] "sample-data.csv" "script.R"
```

To see all the files in your subdirectories, too, use list.files(recursive=T).

A possible "gotcha" of list.files is that it ignores hidden files—typically, any file whose name begins with a period. If you don't see the file you expected to see, try setting all.files=TRUE:

```
> list.files(all.files=TRUE)
```

See Also

R has other handy functions for working with files; see help(files).

4.5 Dealing with "Cannot Open File" in Windows

Problem

You are running R on Windows, and you are using file names such as *C:\data\sample.txt*. R says it cannot open the file, but you know the file does exist.

Solution

The backslashes in the file path are causing trouble. You can solve this problem in one of two ways:

- Change the backslashes to forward slashes: "C:/data/sample.txt".
- Double the backslashes: "C:\\data\\sample.txt".

Discussion

When you open a file in R, you give the file name as a character string. Problems arise when the name contains backslashes (\) because backslashes have a special meaning inside strings. You'll probably get something like this:

```
> samp <- read.csv("C:\Data\sample-data.csv")
Error in file(file, "rt") : cannot open the connection
In addition: Warning messages:
1: '\D' is an unrecognized escape in a character string
```

```
2: '\s' is an unrecognized escape in a character string
3: unrecognized escapes removed from "C:\Data\sample-data.csv"
4: In file(file, "rt") :
   cannot open file 'C:Datasample-data.csv': No such file or directory
```

R escapes every character that follows a backslash and then removes the backslashes. That leaves a meaningless file path, such as *C:Datasample-data.csv* in this example.

The simple solution is to use forward slashes instead of backslashes. R leaves the forward slashes alone, and Windows treats them just like backslashes. Problem solved:

```
> samp <- read.csv("C:/Data/sample-data.csv")
>
```

An alternative solution is to double the backslashes, since R replaces two consecutive backslashes with a single backslash:

```
> samp <- read.csv("C:\\Data\\sample-data.csv")
>
```

4.6 Reading Fixed-Width Records

Problem

You are reading data from a file of fixed-width records: records whose data items occur at fixed boundaries.

Solution

Read the file using the read.fwf function. The main arguments are the file name and the widths of the fields:

```
> records <- read.fwf("filename", widths=c(w₁, w₂, ..., wₙ))
```

Discussion

Suppose we want to read an entire file of fixed-width records, such as *fixed-width.txt*, shown here:

```
Fisher     R.A.       1890 1962
Pearson    Karl       1857 1936
Cox        Gertrude   1900 1978
Yates      Frank      1902 1994
Smith      Kirstine   1878 1939
```

We need to know the column widths. In this case the columns are: last name, 10 characters; first name, 10 characters; year of birth, 4 characters; and year of death, 4 characters. In between the two last columns is a 1-character space. We can read the file this way:

```
> records <- read.fwf("fixed-width.txt", widths=c(10,10,4,-1,4))
```

The `-1` in the `widths` argument says there is a one-character column that should be ignored. The result of `read.fwf` is a data frame:

```
> records
        V1          V2    V3   V4
1 Fisher    R.A.        1890 1962
2 Pearson   Karl        1857 1936
3 Cox       Gertrude    1900 1978
4 Yates     Frank       1902 1994
5 Smith     Kirstine    1878 1939
```

Note that R supplied some funky, synthetic column names. We can override that default by using a `col.names` argument:

```
> records <- read.fwf("fixed-width.txt", widths=c(10,10,4,-1,4),
+                      col.names=c("Last","First","Born","Died"))
> records
        Last    First Born Died
1 Fisher    R.A.        1890 1962
2 Pearson   Karl        1857 1936
3 Cox       Gertrude    1900 1978
4 Yates     Frank       1902 1994
5 Smith     Kirstine    1878 1939
```

`read.fwf` interprets nonnumeric data as a factor (categorical variable) by default. For instance, the Last and First columns just displayed were interpreted as factors. Set `stringsAsFactors=FALSE` to have the function read them as character strings.

The `read.fwf` function has many bells and whistles that may be useful for reading your data files. It also shares many bells and whistles with the `read.table` function. I suggest reading the help pages for both functions.

See Also

See Recipe 4.7 for more discussion of reading text files.

4.7 Reading Tabular Data Files

Problem

You want to read a text file that contains a table of data.

Solution

Use the `read.table` function, which returns a data frame:

```
> dfrm <- read.table("filename")
```

Discussion

Tabular data files are quite common. They are text files with a simple format:

- Each line contains one record.
- Within each record, fields (items) are separated by a one-character delimiter, such as a space, tab, colon, or comma.
- Each record contains the same number of fields.

This format is more free-form than the fixed-width format because fields needn't be aligned by position. Here is the data file of Recipe 4.6 in tabular format, using a space character between fields:

```
Fisher R.A. 1890 1962
Pearson Karl 1857 1936
Cox Gertrude 1900 1978
Yates Frank 1902 1994
Smith Kirstine 1878 1939
```

The read.table function is built to read this file. By default, it assumes the data fields are separated by white space (blanks or tabs):

```
> dfrm <- read.table("statisticians.txt")
> print(dfrm)
        V1       V2   V3   V4
1  Fisher      R.A. 1890 1962
2 Pearson      Karl 1857 1936
3     Cox  Gertrude 1900 1978
4   Yates     Frank 1902 1994
5   Smith  Kirstine 1878 1939
```

If your file uses a separator other than white space, specify it using the sep parameter. If our file used colon (:) as the field separator, we would read it this way:

```
> dfrm <- read.table("statisticians.txt", sep=":")
```

You cannot tell from the printed output, but read.table interpreted the first and last names as factors, not strings. We see that by checking the class of the resulting column:

```
> class(dfrm$V1)
[1] "factor"
```

To prevent read.table from interpreting character strings as factors, set the stringsAsFactors parameter to FALSE:

```
> dfrm <- read.table("statisticians.txt", stringsAsFactor=FALSE)
> class(dfrm$V1)
[1] "character"
```

Now the class of the first column is character, not factor.

If any field contains the string "NA", then read.table assumes that the value is missing and converts it to NA. Your data file might employ a different string to signal missing values, in which case use the na.strings parameter. The SAS convention, for example, is that missing values are signaled by a single period (.). We can read such data files in this way:

```
> dfrm <- read.table("filename.txt", na.strings=".")
```

I am a huge fan of *self-describing data*: data files which describe their own contents. (A computer scientist would say the file contains its own *metadata*.) The `read.table` function has two features that support this characteristic. First, you can include a *header line* at the top of your file that gives names to the columns. The line contains one name for every column, and it uses the same field separator as the data. Here is our data file with a header line that names the columns:

```
lastname firstname born died
Fisher R.A. 1890 1962
Pearson Karl 1857 1936
Cox Gertrude 1900 1978
Yates Frank 1902 1994
Smith Kirstine 1878 1939
```

Now we can tell `read.table` that our file contains a header line, and it will use the column names when it builds the data frame:

```
> dfrm <- read.table("statisticians.txt", header=TRUE, stringsAsFactor=FALSE)
> print(dfrm)
  lastname firstname born died
1   Fisher      R.A. 1890 1962
2  Pearson      Karl 1857 1936
3      Cox  Gertrude 1900 1978
4    Yates     Frank 1902 1994
5    Smith  Kirstine 1878 1939
```

The second feature of `read.table` is *comment lines*. Any line that begins with a pound sign (#) is ignored, so you can put comments on those lines:

```
# This is a data file of famous statisticians.
# Last edited on 1994-06-18
lastname firstname born died
Fisher R.A. 1890 1962
Pearson Karl 1857 1936
Cox Gertrude 1900 1978
Yates Frank 1902 1994
Smith Kirstine 1878 1939
```

`read.table` has many parameters for controlling how it reads and interprets the input file. See the help page for details.

See Also

If your data items are separated by commas, see Recipe 4.8 for reading a CSV file.

4.8 Reading from CSV Files

Problem

You want to read data from a comma-separated values (CSV) file.

Solution

The `read.csv` function can read CSV files. If your CSV file has a header line, use this:

```
> tbl <- read.csv("filename")
```

If your CSV file does not contain a header line, set the `header` option to `FALSE`:

```
> tbl <- read.csv("filename", header=FALSE)
```

Discussion

The CSV file format is popular because many programs can import and export data in that format. Such programs include R, Excel, other spreadsheet programs, many database managers, and most statistical packages. It is a flat file of tabular data, where each line in the file is a row of data, and each row contains data items separated by commas. Here is a very simple CSV file with three rows and three columns (the first line is a *header line* that contains the column names, also separated by commas):

```
label,lbound,ubound
low,0,0.674
mid,0.674,1.64
high,1.64,2.33
```

The `read.csv` file reads the data and creates a data frame, which is the usual R representation for tabular data. The function assumes that your file has a header line unless told otherwise:

```
> tbl <- read.csv("table-data.csv")
> tbl
  label lbound ubound
1   low  0.000  0.674
2   mid  0.674  1.640
3  high  1.640  2.330
```

Observe that `read.csv` took the column names from the header line for the data frame. If the file did not contain a header, then we would specify `header=FALSE` and R would synthesize column names for us (`V1`, `V2`, and `V3` in this case):

```
> tbl <- read.csv("table-data-with-no-header.csv", header=FALSE)
> tbl
    V1    V2    V3
1  low 0.000 0.674
2  mid 0.674 1.640
3 high 1.640 2.330
```

A good feature of `read.csv` is that is automatically interprets nonnumeric data as a factor (categorical variable), which is often what you want since after all this is a statistical package, not Perl. The `label` variable in the `tbl` data frame just shown is actually a factor, not a character variable. You see that by inspecting the structure of `tbl`:

```
> str(tbl)
'data.frame':   3 obs. of  3 variables:
 $ label : Factor w/ 3 levels "high","low","mid": 2 3 1
 $ lbound: num  0 0.674 1.64
 $ ubound: num  0.674 1.64 2.33
```

Sometimes you really want your data interpreted as strings, not as a factor. In that case, set the as.is parameter to TRUE; this indicates that R should not interpret nonnumeric data as a factor:

```
> tbl <- read.csv("table-data.csv", as.is=TRUE)
> str(tbl)
'data.frame':   3 obs. of  3 variables:
 $ label : chr  "low" "mid" "high"
 $ lbound: num  0 0.674 1.64
 $ ubound: num  0.674 1.64 2.33
```

Notice that the label variable now has character-string values and is no longer a factor.

Another useful feature is that input lines starting with a pound sign (#) are ignored, which lets you embed comments in your data file. Disable this feature by specifying comment.char="".

The read.csv function has many useful bells and whistles. These include the ability to skip leading lines in the input file, control the conversion of individual columns, fill out short rows, limit the number of lines, and control the quoting of strings. See the R help page for details.

See Also

See Recipe 4.9. See the R help page for read.table, which is the basis for read.csv.

4.9 Writing to CSV Files

Problem

You want to save a matrix or data frame in a file using the comma-separated values format.

Solution

The write.csv function can write a CSV file:

```
> write.csv(x, file="filename", row.names=FALSE)
```

Discussion

The write.csv function writes tabular data to an ASCII file in CSV format. Each row of data creates one line in the file, with data items separated by commas (,):

```
> print(tbl)
  label lbound ubound
```

```
1    low  0.000  0.674
2    mid  0.674  1.640
3   high  1.640  2.330
> write.csv(tbl, file="table-data.csv", row.names=T)
```

This example creates a file called `table-data.csv` in the current working directory. The file looks like this:

```
"label","lbound","ubound"
"low",0,0.674
"mid",0.674,1.64
"high",1.64,2.33
```

Notice that the function writes a column header line by default. Set `col.names=FALSE` to change that.

If we do not specify `row.names=FALSE`, the function prepends each row with a label taken from the `row.names` attribute of your data. If your data doesn't have row names then the function just uses the row numbers, which creates a CSV file like this:

```
"","label","lbound","ubound"
"1","low",0,0.674
"2","mid",0.674,1.64
"3","high",1.64,2.33
```

I rarely want row labels in my CSV files, which is why I recommend setting `row.names=FALSE`.

The function is intentionally inflexible. You cannot easily change the defaults because it really, really wants to write files in a valid CSV format. Use the `write.table` function to save your tabular data in other formats.

A sad limitation of `write.csv` is that it cannot append lines to a file. Use `write.table` instead.

See Also

See Recipe 3.1 for more about the current working directory and Recipe 4.14 for other ways to save data to files.

4.10 Reading Tabular or CSV Data from the Web

Problem

You want to read data directly from the Web into your R workspace.

Solution

Use the `read.csv`, `read.table`, and `scan` functions, but substitute a URL for a file name. The functions will read directly from the remote server:

```
> tbl <- read.csv("http://www.example.com/download/data.csv")
```

You can also open a connection using the URL and then read from the connection, which may be preferable for complicated files.

Discussion

The Web is a gold mine of data. You could download the data into a file and then read the file into R, but it's more convenient to read directly from the Web. Give the URL to `read.csv`, `read.table`, or `scan` (depending upon the format of the data), and the data will be downloaded and parsed for you. No fuss, no muss.

Aside from using a URL, this recipe is just like reading from a CSV file (Recipe 4.8) or a complex file (Recipe 4.12), so all the comments in those recipes apply here, too.

Remember that URLs work for FTP servers, not just HTTP servers. This means that R can also read data from FTP sites using URLs:

```
> tbl <- read.table("ftp://ftp.example.com/download/data.txt")
```

See Also

See Recipes 4.8 and 4.12.

4.11 Reading Data from HTML Tables

Problem

You want to read data from an HTML table on the Web.

Solution

Use the `readHTMLTable` function in the `XML` package. To read all tables on the page, simply give the URL:

```
> library(XML)
> url <- 'http://www.example.com/data/table.html'
> tbls <- readHTMLTable(url)
```

To read only specific tables, use the `which` parameter. This example reads the third table on the page:

```
> tbl <- readHTMLTable(url, which=3)
```

Discussion

Web pages can contain several HTML tables. Calling `readHTMLTable(url)` reads all tables on the page and returns them in a list. This can be useful for exploring a page, but it's annoying if you want just one specific table. In that case, use `which=n` to select the desired table. You'll get only the *n*th table.

The following example, which is taken from the help page for readHTMLTable, loads all tables from the Wikipedia page entitled "World population":

```
> library(XML)
> url <- 'http://en.wikipedia.org/wiki/World_population'
> tbls <- readHTMLTable(url)
```

As it turns out, that page contains 17 tables:

```
> length(tbls)
[1] 17
```

In this example we care only about the third table (which lists the largest populations by country), so we specify which=3:

```
> tbl <- readHTMLTable(url, which=3)
```

In that table, columns 2 and 3 contain the country name and population, respectively:

```
> tbl[,c(2,3)]
                   Country / Territory   Population
1  Â People's Republic of China[44]  1,338,460,000
2                          Â India  1,182,800,000
3                  Â United States    309,659,000
4                     Â Indonesia    231,369,500
5                        Â Brazil    193,152,000
6                      Â Pakistan    169,928,500
7                    Â Bangladesh    162,221,000
8                       Â Nigeria    154,729,000
9                        Â Russia    141,927,297
10                        Â Japan    127,530,000
11                       Â Mexico    107,550,697
12                   Â Philippines     92,226,600
13                      Â Vietnam     85,789,573
14                      Â Germany     81,882,342
15                     Â Ethiopia     79,221,000
16                        Â Egypt     78,459,000
```

Right away, we can see problems with the data: the country names have some funky Unicode character stuck to the front. I don't know why; it probably has something to do with formatting the Wikipedia page. Also, the name of the People's Republic of China has "[44]" appended. On the Wikipedia website, that was a footnote reference, but now it's just a bit of unwanted text. Adding insult to injury, the population numbers have embedded commas, so you cannot easily convert them to raw numbers. All these problems can be solved by some string processing, but each problem adds at least one more step to the process.

This illustrates the main obstacle to reading HTML tables. HTML was designed for presenting information to people, not to computers. When you "scrape" information off an HTML page, you get stuff that's useful to people but annoying to computers. If you ever have a choice, choose instead a computer-oriented data representation such as XML, JSON, or CSV.

The readHTMLTable function is part of the XML package, which (by necessity) is large and complex. The XML package depends on a software library called libxml2, which you will need to obtain and install first. On Linux, you will also need the Linux package xml2-config, which is necessary for building the R package.

See Also

See Recipe 3.9 for downloading and installing packages such as the XML package.

4.12 Reading Files with a Complex Structure

Problem

You are reading data from a file that has a complex or irregular structure.

Solution

- Use the readLines function to read individual lines; then process them as strings to extract data items.
- Alternatively, use the scan function to read individual tokens and use the argument what to describe the stream of tokens in your file. The function can convert tokens into data and then assemble the data into records.

Discussion

Life would be simple and beautiful if all our data files were organized into neat tables with cleanly delimited data. We could read those files using read.table and get on with living.

Dream on.

You will eventually encounter a funky file format, and your job—no matter how painful—is to read the file contents into R. The read.table and read.csv functions are line-oriented and probably won't help. However, the readLines and scan functions are useful here because they let you process the individual lines and even tokens of the file.

The readLines function is pretty simple. It reads lines from a file and returns them as a list of character strings:

```
> lines <- readLines("input.txt")
```

You can limit the number of lines by using the n parameter, which gives the number of maximum number of lines to be read:

```
> lines <- readLines("input.txt", n=10)       # Read 10 lines and stop
```

The scan function is much richer. It reads one token at a time and handles it according to your instructions. The first argument is either a filename or a connection (more on connections later). The second argument is called what, and it describes the tokens that scan should expect in the input file. The description is cryptic but quite clever:

what=numeric(0)
> Interpret the next token as a number.

what=integer(0)
> Interpret the next token as an integer.

what=complex(0)
> Interpret the next token as complex number.

what=character(0)
> Interpret the next token as a character string.

what=logical(0)
> Interpret the next token as a logical value.

The scan function will apply the given pattern repeatedly until all data is read.

Suppose your file is simply a sequence of numbers, like this:

```
2355.09 2246.73 1738.74 1841.01 2027.85
```

Use what=numeric(0) to say, "My file is a sequence of tokens, each of which is a number":

```
> singles <- scan("singles.txt", what=numeric(0))
Read 5 items
> singles
[1] 2355.09 2246.73 1738.74 1841.01 2027.85
```

A key feature of scan is that the what can be a list containing several token types. The scan function will assume your file is a repeating sequence of those types. Suppose your file contains triplets of data, like this:

```
15-Oct-87 2439.78 2345.63 16-Oct-87 2396.21 2,207.73
19-Oct-87 2164.16 1677.55 20-Oct-87 2067.47 1,616.21
21-Oct-87 2081.07 1951.76
```

Use a list to tell scan that it should expect a repeating, three-token sequence:

```
> triples <- scan("triples.txt", what=list(character(0),numeric(0),numeric(0)))
```

Give names to the list elements, and scan will assign those names to the data:

```
> triples <- scan("triples.txt",
+                  what=list(date=character(0), high=numeric(0), low=numeric(0)))
Read 5 records
> triples
$date
[1] "15-Oct-87" "16-Oct-87" "19-Oct-87" "20-Oct-87" "21-Oct-87"

$high
[1] 2439.78 2396.21 2164.16 2067.47 2081.07
```

```
$low
[1] 2345.63 2207.73 1677.55 1616.21 1951.76
```

The scan function has many bells and whistles, but the following are especially useful:

n=*number*
 Stop after reading this many tokens. (Default: stop at end of file.)

nlines=*number*
 Stop after reading this many input lines. (Default: stop at end of file.)

skip=*number*
 Number of input lines to skip before reading data.

na.strings=*list*
 A list of strings to be interpreted as NA.

An Example

Let's use this recipe to read a dataset from StatLib, the repository of statistical data and software maintained by Carnegie Mellon University. Jeff Witmer contributed a dataset called wseries that shows the pattern of wins and losses for every World Series since 1903. The dataset is stored in an ASCII file with 35 lines of comments followed by 23 lines of data. The data itself looks like this:

```
1903   LWLlwwwW    1927   wwWW    1950   wwWW    1973   WLwllWW
1905   wLwWW       1928   WWww    1951   LWlwwW   1974   wlWWW
1906   wLwLwW      1929   wwLWW   1952   lwLWLww  1975   lwWLWlw
1907   WWww        1930   WWllwW  1953   WWllwW   1976   WWww
1908   wWLww       1931   LWwlwLW 1954   WWww     1977   WLwwlW

.
. (etc.)
.
```

The data is encoded as follows: L = loss at home, l = loss on the road, W = win at home, w = win on the road. The data appears in column order, not row order, which complicates our lives a bit.

Here is the R code for reading the raw data:

```
# Read the wseries dataset:
#     - Skip the first 35 lines
#     - Then read 23 lines of data
#     - The data occurs in pairs: a year and a pattern (char string)
#
world.series <- scan("http://lib.stat.cmu.edu/datasets/wseries",
                     skip = 35,
                     nlines = 23,
                     what = list(year = integer(0),
                                 pattern = character(0)),
                     )
```

The scan function returns a list, so we get a list with two elements: year and pattern. The function reads from left to right, but the dataset is organized by columns and so the years appear in a strange order:

```
> world.series$year
 [1] 1903 1927 1950 1973 1905 1928 1951 1974 1906 1929 1952
[12] 1975 1907 1930 1953 1976 1908 1931 1954 1977 1909 1932

.
. (etc.)
.
```

We can fix that by sorting the list elements according to year:

```
> perm <- order(world.series$year)
> world.series <- list(year    = world.series$year[perm],
+                      pattern = world.series$pattern[perm])
```

Now the data appears in chronological order:

```
> world.series$year
 [1] 1903 1905 1906 1907 1908 1909 1910 1911 1912 1913 1914
[12] 1915 1916 1917 1918 1919 1920 1921 1922 1923 1924 1925

.
. (etc.)
.

> world.series$pattern
 [1] "LWLlwwwW" "wLwWW"    "wLwLwW"   "WWww"     "wWLww"
 [6] "WLwlWlw"  "WWwlw"    "lWwWlW"   "wLwWlLW"  "wLwWw"
[11] "wwWW"     "lwWWw"    "WWlwW"    "WWllWw"   "wlwWLW"
[16] "WWlwwLLw" "wllWWWW"  "LlWwLwWw" "WWwW"     "LwLwWw"

.
. (etc.)
.
```

4.13 Reading from MySQL Databases

Problem

You want access to data stored in a MySQL database.

Solution

1. Install the RMySQL package on your computer.
2. Open a database connection using the dbConnect function.
3. Use dbGetQuery to initiate a SELECT and return the result sets.
4. Use dbDisconnect to terminate the database connection when you are done.

Discussion

This recipe requires that the RMySQL package be installed on your computer. That package requires, in turn, the MySQL client software. If the MySQL client software is not already installed and configured, consult the MySQL documentation or your system administrator.

Use the dbConnect function to establish a connection to the MySQL database. It returns a connection object which is used in subsequent calls to RMySQL functions:

```
library(RMySQL)
con <- dbConnect(MySQL(), user="userid", password="pswd",
                host="hostname", client.flag=CLIENT_MULTI_RESULTS)
```

Setting client.flag=CLIENT_MULTI_RESULTS is necessary to correctly handle multiple result sets. Even if your queries return a single result set, you must set client.flag this way because MySQL might include additional status result sets after your data.

The username, password, and host parameters are the same parameters used for accessing MySQL through the *mysql* client program. The example given here shows them hard-coded into the dbConnect call. Actually, that is an ill-advised practice. It puts your password out in the open, creating a security problem. It also creates a major headache whenever your password or host change, requiring you to hunt down the hard-coded values. I strongly recommend using the security mechanism of MySQL instead. Put those three parameters into your MySQL configuration file, which is *$HOME/.my.cnf* on Unix and *C:\my.cnf* on Windows. Make sure the file is unreadable by anyone except you. The file is delimited into sections with markers such as [client]. Put the parameters into the [client] section, so that your config file will contain something like this:

```
[client]
user = userid
password = password
host = hostname
```

Once the parameters are defined in the config file, you no longer need to supply them in the dbConnect call, which then becomes much simpler:

```
con <- dbConnect(MySQL(), client.flag=CLIENT_MULTI_RESULTS)
```

Use the dbGetQuery function to submit your SQL to the database and read the result sets. Doing so requires an open database connection:

```
sql <- "SELECT * from SurveyResults WHERE City = 'Chicago'"
rows <- dbGetQuery(con, sql)
```

You will need to construct your own SQL query, of course; this is just an example. You are not restricted to SELECT statements. Any SQL that generates a result set is OK. I generally use CALL statements, for example, because all my SQL is encapsulated in stored procedures and those stored procedures contain embedded SELECT statements.

Using dbGetQuery is convenient because it packages the result set into a data frame and returns the data frame. This is the perfect representation of an SQL result set. The result set is a tabular data structure of rows and columns, and so is a data frame. The result set's columns have names given by the SQL SELECT statement, and R uses them for naming the columns of the data frame.

After the first result set of data, MySQL can return a second result set containing status information. You can choose to inspect the status or ignore it, but you must read it. Otherwise, MySQL will complain that there are unprocessed result sets and then halt. So call dbNextResult if necessary:

```
if (dbMoreResults(con)) dbNextResult(con)
```

Call dbGetQuery repeatedly to perform multiple queries, checking for the result status after each call (and reading it, if necessary). When you are done, close the database connection using dbDisconnect:

```
dbDisconnect(con)
```

Here is a complete session that reads and prints three rows from my database of stock prices. The query selects the price of IBM stock for the last three days of 2008. It assumes that the username, password, and host are defined in the *my.cnf* file:

```
> con <- dbConnect(MySQL(), client.flag=CLIENT_MULTI_RESULTS)
> sql <- paste("select * from DailyBar where Symbol = 'IBM'",
+              "and Day between '2008-12-29' and '2008-12-31'")
> rows <- dbGetQuery(con, sql)
> if (dbMoreResults(con)) dbNextResults(con)
> print(rows)
  Symbol        Day       Next OpenPx HighPx LowPx ClosePx AdjClosePx
1    IBM 2008-12-29 2008-12-30  81.72  81.72 79.68   81.25      81.25
2    IBM 2008-12-30 2008-12-31  81.83  83.64 81.52   83.55      83.55
3    IBM 2008-12-31 2009-01-02  83.50  85.00 83.50   84.16      84.16
  HistClosePx  Volume OpenInt
1       81.25 6062600      NA
2       83.55 5774400      NA
3       84.16 6667700      NA
> dbDisconnect(con)
[1] TRUE
```

See Also

See Recipe 3.9 and the documentation for RMySQL, which contains more details about configuring and using the package.

R can read from several other RDBMS systems, including Oracle, Sybase, PostgreSQL, and SQLite. For more information, see the *R Data Import/Export* guide, which is supplied with the base distribution (Recipe 1.6) and is also available on CRAN at *http:// cran.r-project.org/doc/manuals/R-data.pdf*.

4.14 Saving and Transporting Objects

Problem

You want to store one or more R objects in a file for later use, or you want to copy an R object from one machine to another.

Solution

Write the objects to a file using the **save** function:

```
> save(myData, file="myData.RData")
```

Read them back using the `load` function, either on your computer or on any platform that supports R:

```
> load("myData.RData")
```

The **save** function writes binary data. To save in an ASCII format, use `dput` or `dump` instead:

```
> dput(myData, file="myData.txt")
> dump("myData", file="myData.txt")          # Note quotes around variable name
```

Discussion

I normally save my data in my workspace, but sometimes I need to save data outside my workspace. I may have a large, complicated data object that I want to load into other workspaces, or I may want to move R objects between my Linux box and my Windows box. The `load` and **save** functions let me do all this: **save** will store the object in a file that is portable across machines, and `load` can read those files.

When you run `load`, it does not return your data per se; rather, it creates variables in your workspace, loads your data into those variables, and then returns the names of the variables (in a list). The first time I used `load`, I did this:

```
> myData <- load("myFile.RData")     # Achtung! Might not do what you think
```

I was extremely puzzled because myData did not contain my data at all and because my variables had mysteriously appeared in my workspace. Eventually, I broke down and read the documentation for `load`, which explained everything.

The **save** function writes in a binary format to keep the file small. Sometimes you want an ASCII format instead. When you submit a question to a mailing list, for example, including an ASCII dump of the data lets others re-create your problem. In such cases use `dput` or `dump`, which write an ASCII representation.

Be careful when you save and load objects created by a particular R package. When you load the objects, R does not automatically load the required packages, too, so it will not "understand" the object unless you previously loaded the package yourself. For instance, suppose you have an object called z created by the zoo package, and

suppose we save the object in a file called *z.RData*. The following sequence of functions will create some confusion:

```
> load("z.RData")      # Create and populate the z variable
> plot(z)              # Does not plot what we expected: zoo pkg not loaded
```

We should have loaded the zoo package *before* printing or plotting any zoo objects, like this:

```
> library(zoo)         # Load the zoo package into memory
> load("z.RData")      # Create and populate the z variable
> plot(z)              # Ahhh. Now plotting works correctly
```

Data Structures

Introduction

You can get pretty far in R just using vectors. That's what Chapter 2 is all about. This chapter moves beyond vectors to recipes for matrices, lists, factors, and data frames. If you have preconceptions about data structures, I suggest you put them aside. R does data structures differently.

If you want to study the technical aspects of R's data structures, I suggest reading *R in a Nutshell* (O'Reilly) and the *R Language Definition*. My notes here are more informal. These are things I wish I'd known when I started using R.

Vectors

Here are some key properties of vectors:

Vectors are homogeneous

> All elements of a vector must have the same type or, in R terminology, the same mode.

Vectors can be indexed by position

> So v[2] refers to the second element of v.

Vectors can be indexed by multiple positions, returning a subvector

> So v[c(2,3)] is a subvector of v that consists of the second and third elements.

Vector elements can have names

> Vectors have a names property, the same length as the vector itself, that gives names to the elements:

```
> v <- c(10, 20, 30)
> names(v) <- c("Moe", "Larry", "Curly")
> print(v)
  Moe Larry Curly
   10    20    30
```

If vector elements have names then you can select them by name

Continuing the previous example:

```
> v["Larry"]
Larry
   20
```

Lists

Lists are heterogeneous

Lists can contain elements of different types; in R terminology, list elements may have different modes. Lists can even contain other structured objects, such as lists and data frames; this allows you to create recursive data structures.

Lists can be indexed by position

So `lst[[2]]` refers to the second element of `lst`. Note the double square brackets.

Lists let you extract sublists

So `lst[c(2,3)]` is a sublist of `lst` that consists of the second and third elements. Note the single square brackets.

List elements can have names

Both `lst[["Moe"]]` and `lst$Moe` refer to the element named "Moe".

Since lists are heterogeneous and since their elements can be retrieved by name, a list is like a *dictionary* or *hash* or *lookup table* in other programming languages (Recipe 5.9). What's surprising (and cool) is that in R, unlike most of those other programming languages, lists can also be indexed by position.

Mode: Physical Type

In R, every object has a *mode*, which indicates how it is stored in memory: as a number, as a character string, as a list of pointers to other objects, as a function, and so forth:

Object	Example	Mode
Number	3.1415	numeric
Vector of numbers	c(2.7.182, 3.1415)	numeric
Character string	"Moe"	character
Vector of character strings	c("Moe", "Larry", "Curly")	character
Factor	factor(c("NY", "CA", "IL"))	numeric
List	list("Moe", "Larry", "Curly")	list
Data frame	data.frame(x=1:3, y=c("NY", "CA", "IL"))	list
Function	print	function

The `mode` function gives us this information:

```
> mode(3.1415)                          # Mode of a number
[1] "numeric"
> mode(c(2.7182, 3.1415))               # Mode of a vector of numbers
[1] "numeric"
> mode("Moe")                           # Mode of a character string
[1] "character"
> mode(list("Moe","Larry","Curly"))    # Mode of a list
[1] "list"
```

A critical difference between a vector and a list can be summed up this way:

- In a vector, all elements must have the same mode.
- In a list, the elements can have different modes.

Class: Abstract Type

In R, every object also has a *class*, which defines its abstract type. The terminology is borrowed from object-oriented programming. A single number could represent many different things: a distance, a point in time, a weight. All those objects have a mode of "numeric" because they are stored as a number; but they could have different classes to indicate their interpretation.

For example, a Date object consists of a single number:

```
> d <- as.Date("2010-03-15")
> mode(d)
[1] "numeric"
> length(d)
[1] 1
```

But it has a class of Date, telling us how to interpret that number; namely, as the number of days since January 1, 1970:

```
> class(d)
[1] "Date"
```

R uses an object's class to decide how to process the object. For example, the generic function print has specialized versions (called *methods*) for printing objects according to their class: data.frame, Date, lm, and so forth. When you print an object, R calls the appropriate print function according to the object's class.

Scalars

The quirky thing about scalars is their relationship to vectors. In some software, scalars and vectors are two different things. In R, they are the same thing: a scalar is simply a vector that contains exactly one element. In this book I often use the term "scalar", but that's just shorthand for "vector with one element."

Consider the built-in constant pi. It is a scalar:

```
> pi
[1] 3.141593
```

Since a scalar is a one-element vector, you can use vector functions on `pi`:

```
> length(pi)
[1] 1
```

You can index it. The first (and only) element is π, of course:

```
> pi[1]
[1] 3.141593
```

If you ask for the second element, there is none:

```
> pi[2]
[1] NA
```

Matrices

In R, a matrix is just a vector that has dimensions. It may seem strange at first, but you can transform a vector into a matrix simply by giving it dimensions.

A vector has an attribute called `dim`, which is initially `NULL`, as shown here:

```
> A <- 1:6
> dim(A)
NULL
> print(A)
[1] 1 2 3 4 5 6
```

We give dimensions to the vector when we set its `dim` attribute. Watch what happens when we set our vector dimensions to 2×3 and print it:

```
> dim(A) <- c(2,3)
> print(A)
     [,1] [,2] [,3]
[1,]    1    3    5
[2,]    2    4    6
```

Voilà! The vector was reshaped into a 2×3 matrix.

A matrix can be created from a list, too. Like a vector, a list has a `dim` attribute, which is initially `NULL`:

```
> B <- list(1,2,3,4,5,6)
> dim(B)
NULL
```

If we set the `dim` attribute, it gives the list a shape:

```
> dim(B) <- c(2,3)
> print(B)
     [,1] [,2] [,3]
[1,] 1    3    5
[2,] 2    4    6
```

Voilà! We have turned this list into a 2×3 matrix.

Arrays

The discussion of matrices can be generalized to 3-dimensional or even *n*-dimensional structures: just assign more dimensions to the underlying vector (or list). The following example creates a 3-dimensional array with dimensions 2 × 3 × 2:

```
> D <- 1:12
> dim(D) <- c(2,3,2)
> print(D)
, , 1

     [,1] [,2] [,3]
[1,]    1    3    5
[2,]    2    4    6

, , 2

     [,1] [,2] [,3]
[1,]    7    9   11
[2,]    8   10   12
```

Note that R prints one "slice" of the structure at a time, since it's not possible to print a 3-dimensional structure on a 2-dimensional medium.

Matrices Made from Lists

It strikes me as very odd that we can turn a list into a matrix just by giving the list a `dim` attribute. But wait; it gets stranger.

Recall that a list can be heterogeneous (mixed modes). We can start with a heterogeneous list, give it dimensions, and thus create a heterogeneous matrix. This code snippet creates a matrix that is a mix of numeric and character data:

```
> C <- list(1, 2, 3, "X", "Y", "Z")
> dim(C) <- c(2,3)
> print(C)
     [,1] [,2] [,3]
[1,] 1    3    "Y"
[2,] 2    "X"  "Z"
```

To me this is strange because I ordinarily assume a matrix is purely numeric, not mixed. R is not that restrictive.

The possibility of a heterogeneous matrix may seem powerful and strangely fascinating. However, it creates problems when you are doing normal, day-to-day stuff with matrices. For example, what happens when the matrix C (above) is used in matrix multiplication? What happens if it is converted to a data frame? The answer is that odd things happen.

In this book, I generally ignore the pathological case of a heterogeneous matrix. I assume you've got simple, vanilla matrices. Some recipes involving matrices may work oddly (or not at all) if your matrix contains mixed data. Converting such a matrix to a vector or data frame, for instance, can be problematic (Recipe 5.33).

Factors

A factor looks like a vector, but it has special properties. R keeps track of the unique values in a vector, and each unique value is called a *level* of the associated factor. R uses a compact representation for factors, which makes them efficient for storage in data frames. In other programming languages, a factor would be represented by a *vector of enumerated values*.

There are two key uses for factors:

Categorical variables
> A factor can represent a categorical variable. Categorical variables are used in contingency tables, linear regression, analysis of variance (ANOVA), logistic regression, and many other areas.

Grouping
> This is a technique for labeling or tagging your data items according to their group. See the "Introduction" to Chapter 6.

Data Frames

A data frame is powerful and flexible structure. Most serious R applications involve data frames. A data frame is intended to mimic a dataset, such as one you might encounter in SAS or SPSS.

A data frame is a tabular (rectangular) data structure, which means that it has rows and columns. It is not implemented by a matrix, however. Rather, a data frame is a list:

- The elements of the list are vectors and/or factors.*
- Those vectors and factors are the columns of the data frame.
- The vectors and factors must all have the same length; in other words, all columns must have the same height.
- The equal-height columns give a rectangular shape to the data frame.
- The columns must have names.

Because a data frame is both a list and a rectangular structure, R provides two different paradigms for accessing its contents:

- You can use list operators to extract columns from a data frame, such as dfrm[i], dfrm[[i]], or dfrm$*name*.
- You can use matrix-like notation, such as dfrm[i,j], dfrm[i,], or dfrm[,j].

Your perception of a data frame likely depends on your background:

* A data frame can be built from a mixture of vectors, factors, and matrices. The columns of the matrices become columns in the data frame. The number of *rows* in each matrix must match the *length* of the vectors and factors. In other words, all elements of a data frame must have the same *height*.

To a statistician

A data frame is a table of observations. Each row contains one observation. Each observation must contain the same variables. These variables are called columns, and you can refer to them by name. You can also refer to the contents by row number and column number, just as with a matrix.

To a SQL programmer

A data frame is a table. The table resides entirely in memory, but you can save it to a flat file and restore it later. You needn't declare the column types because R figures that out for you.

To an Excel user

A data frame is like a worksheet, or perhaps a range within a worksheet. It is more restrictive, however, in that each column has a type.

To an SAS user

A data frame is like a SAS dataset for which all the data resides in memory. R can read and write the data frame to disk, but the data frame must be in memory while R is processing it.

To an R programmer

A data frame is a hybrid data structure, part matrix and part list. A column can contain numbers, character strings, or factors but not a mix of them. You can index the data frame just like you index a matrix. The data frame is also a list, where the list elements are the columns, so you can access columns by using list operators.

To a computer scientist

A data frame is a rectangular data structure. The columns are strongly typed, and each column must be numeric values, character strings, or a factor. Columns must have labels; rows may have labels. The table can be indexed by position, column name, and/or row name. It can also be accessed by list operators, in which case R treats the data frame as a list whose elements are the columns of the data frame.

To an executive

You can put names and numbers into a data frame. It's easy! A data frame is like a little database. Your staff will enjoy using data frames.

5.1 Appending Data to a Vector

Problem

You want to append additional data items to a vector.

Solution

Use the vector constructor (c) to construct a vector with the additional data items:

```
> v <- c(v,newItems)
```

For a single item, you can also assign the new item to the next vector element. R will automatically extend the vector:

```
> v[length(v)+1] <- newItem
```

Discussion

If you ask me about appending a data item to a vector, I will suggest that maybe you shouldn't.

 R works best when you think about entire vectors, not single data items. Are you repeatedly appending items to a vector? If so, then you are probably working inside a loop. That's OK for small vectors, but for large vectors your program will run slowly. The memory management in R works poorly when you repeatedly extend a vector by one element. Try to replace that loop with vector-level operations. You'll write less code, and R will run much faster.

Nonetheless, one does occasionally need to append data to vectors. My experiments show that the most efficient way is to create a new vector using the vector constructor (c) to join the old and new data. This works for appending single elements or multiple elements:

```
> v <- c(1,2,3)
> v <- c(v,4)          # Append a single value to v
> v
[1] 1 2 3 4
> w <- c(5,6,7,8)
> v <- c(v,w)          # Append an entire vector to v
> v
[1] 1 2 3 4 5 6 7 8
```

You can also append an item by assigning it to the position past the end of the vector, as shown in the Solution. In fact, R is very liberal about extending vectors. You can assign to any element and R will expand the vector to accommodate your request:

```
> v <- c(1,2,3)        # Create a vector of three elements
> v[10] <- 10          # Assign to the 10th element
> v                    # R extends the vector automatically
 [1]  1  2  3 NA NA NA NA NA NA 10
```

Note that R did not complain about the out-of-bounds subscript. It just extended the vector to the needed length, filling with NA.

R includes an append function that creates a new vector by appending items to an existing vector. However, my experiments show that this function runs more slowly than both the vector constructor and the element assignment.

5.2 Inserting Data into a Vector

Problem

You want to insert one or more data items into a vector.

Solution

Despite its name, the append function inserts data into a vector by using the after parameter, which gives the insertion point for the new item or items:

```
> append(vec, newvalues, after=n)
```

Discussion

The new items will be inserted at the position given by after. This example inserts 99 into the middle of a sequence:

```
> append(1:10, 99, after=5)
 [1]  1  2  3  4  5 99  6  7  8  9 10
```

The special value of after=0 means insert the new items at the head of the vector:

```
> append(1:10, 99, after=0)
 [1] 99  1  2  3  4  5  6  7  8  9 10
```

The comments in Recipe 5.1 apply here, too. If you are inserting single items into a vector, you might be working at the element level when working at the vector level would be easier to code and faster to run.

5.3 Understanding the Recycling Rule

Problem

You want to understand the mysterious Recycling Rule that governs how R handles vectors of unequal length.

Discussion

When you do vector arithmetic, R performs element-by-element operations. That works well when both vectors have the same length: R pairs the elements of the vectors and applies the operation to those pairs.

But what happens when the vectors have unequal lengths?

In that case, R invokes the Recycling Rule. It processes the vector element in pairs, starting at the first elements of both vectors. At a certain point, the shorter vector is exhausted while the longer vector still has unprocessed elements. R returns to the beginning of the shorter vector, "recycling" its elements; continues taking elements from

the longer vector; and completes the operation. It will recycle the shorter-vector elements as often as necessary until the operation is complete.

It's useful to visualize the Recycling Rule. Here is a diagram of two vectors, 1:6 and 1:3:

1:6	1:3
1	1
2	2
3	3
4	
5	
6	

Obviously, the 1:6 vector is longer than the 1:3 vector. If we try to add the vectors using (1:6) + (1:3), it appears that 1:3 has too few elements. However, R recycles the elements of 1:3, pairing the two vectors like this and producing a six-element vector:

1:6	1:3	(1:6) + (1:3)
1	1	2
2	2	4
3	3	6
4	1	5
5	2	7
6	3	9

Here is what you see at the command line:

```
> (1:6) + (1:3)
[1] 2 4 6 5 7 9
```

It's not only vector operations that invoke the Recycling Rule; functions can, too. The cbind function can create column vectors, such as the following column vectors of 1:6 and 1:3. The two column have different heights, of course:

```
> cbind(1:6)
     [,1]
[1,]    1
[2,]    2
[3,]    3
[4,]    4
[5,]    5
[6,]    6
```

```
> cbind(1:3)
     [,1]
[1,]   1
[2,]   2
[3,]   3
```

If we try binding these column vectors together into a two-column matrix, the lengths are mismatched. The 1:3 vector is too short, so cbind invokes the Recycling Rule and recycles the elements of 1:3:

```
> cbind(1:6, 1:3)
     [,1] [,2]
[1,]   1    1
[2,]   2    2
[3,]   3    3
[4,]   4    1
[5,]   5    2
[6,]   6    3
```

If the longer vector's length is not a multiple of the shorter vector's length, R gives a warning. That's good, since the operation is highly suspect and there is likely a bug in your logic:

```
> (1:6) + (1:5)          # Oops! 1:5 is one element too short
[1]  2  4  6  8 10  7
Warning message:
In (1:6) + (1:5) :
  longer object length is not a multiple of shorter object length
```

Once you understand the Recycling Rule, you will realize that operations between a vector and a scalar are simply applications of that rule. In this example, the 10 is recycled repeatedly until the vector addition is complete:

```
> (1:6) + 10
[1] 11 12 13 14 15 16
```

5.4 Creating a Factor (Categorical Variable)

Problem

You have a vector of character strings or integers. You want R to treat them as a factor, which is R's term for a categorical variable.

Solution

The factor function encodes your vector of discrete values into a factor:

```
> f <- factor(v)          # v is a vector of strings or integers
```

If your vector contains only a subset of possible values and not the entire universe, then include a second argument that gives the possible levels of the factor:

```
> f <- factor(v, levels)
```

Discussion

In R, each possible value of a categorical variable is called a *level*. A vector of levels is called a *factor*. Factors fit very cleanly into the vector orientation of R, and they are used in powerful ways for processing data and building statistical models.

Most of the time, converting your categorical data into a factor is a simple matter of calling the `factor` function, which identifies the distinct levels of the categorical data and packs them into a factor:

```
> f <- factor(c("Win","Win","Lose","Tie","Win","Lose"))
> f
[1] Win  Win  Lose Tie  Win  Lose
Levels: Lose Tie Win
```

Notice that when we printed the factor, f, R did not put quotes around the values. They are levels, not strings. Also notice that when we printed the factor, R also displayed the distinct levels below the factor.

If your vector contains only a subset of all the possible levels, then R will have an incomplete picture of the possible levels. Suppose you have a string-valued variable wday that gives the day of the week on which your data was observed:

```
> f <- factor(wday)
> f
 [1] Wed Thu Mon Wed Thu Thu Thu Tue Thu Tue
Levels: Mon Thu Tue Wed
```

R thinks that Monday, Thursday, Tuesday, and Wednesday are the only possible levels. Friday is not listed. Apparently, the lab staff never made observations on Friday, so R does not know that Friday is a possible value. Hence you need to list the possible levels of wday explicitly:

```
> f <- factor(wday, c("Mon","Tue","Wed","Thu","Fri"))
> f
 [1] Wed Thu Mon Wed Thu Thu Thu Tue Thu Tue
Levels: Mon Tue Wed Thu Fri
```

Now R understands that f is a factor with five possible levels. It knows their correct order, too. It originally put Thursday before Tuesday because it assumes alphabetical order by default.[†] The explicit second argument defines the correct order.

In many situations it is not necessary to call `factor` explicitly. When an R function requires a factor, it usually converts your data to a factor automatically. The `table` function, for instance, works only on factors, so it routinely converts its inputs to factors without asking. You must explicitly create a factor variable when you want to specify the full set of levels or when you want to control the ordering of levels.

† More precisely, it orders the names according to your Locale.

See Also

See Recipe 12.6 to create a factor from continuous data.

5.5 Combining Multiple Vectors into One Vector and a Factor

Problem

You have several groups of data, with one vector for each group. You want to combine the vectors into one large vector and simultaneously create a parallel factor that identifies each value's original group.

Solution

Create a list that contains the vectors. Use the stack function to combine the list into a two-column data frame:

```
> comb <- stack(list(v1=v1, v2=v2, v3=v3))     # Combine 3 vectors
```

The data frame's columns are called values and ind. The first column contains the data, and the second column contains the parallel factor.

Discussion

Why in the world would you want to mash all your data into one big vector and a parallel factor? The reason is that many important statistical functions require the data in that format.

Suppose you survey freshmen, sophomores, and juniors regarding their confidence level ("What percentage of the time do you feel confident in school?"). Now you have three vectors, called freshmen, sophomores, and juniors. You want to perform an ANOVA analysis of the differences between the groups. The ANOVA function, aov, requires one vector with the survey results as well as a parallel factor that identifies the group. You can combine the groups using the stack function:

```
> comb <- stack(list(fresh=freshmen, soph=sophomores, jrs=juniors))
> print(comb)
   values   ind
1    0.60 fresh
2    0.35 fresh
3    0.44 fresh
4    0.62 fresh
5    0.60 fresh
6    0.70  soph
7    0.61  soph
8    0.63  soph
9    0.87  soph
10   0.85  soph
11   0.70  soph
12   0.64  soph
```

```
13    0.76    jrs
14    0.71    jrs
15    0.92    jrs
16    0.87    jrs
```

Now you can perform the ANOVA analysis on the two columns:

```
> aov(values ~ ind, data=comb)
```

When building the list we must provide tags for the list elements (the tags are fresh, soph, and jrs in this example). Those tags are required because stack uses them as the levels of the parallel factor.

5.6 Creating a List

Problem

You want to create and populate a list.

Solution

To create a list from individual data items, use the list function:

```
> lst <- list(x, y, z)
```

Discussion

Lists can be quite simple, such as this list of three numbers:

```
> lst <- list(0.5, 0.841, 0.977)
> lst
[[1]]
[1] 0.5

[[2]]
[1] 0.841

[[3]]
[1] 0.977
```

When R prints the list, it identifies each list element by its position ([[1]], [[2]], [[3]]) and prints the element's value (e.g., [1] 0.5) under its position.

More usefully, lists can—unlike vectors—contain elements of different modes (types). Here is an extreme example of a mongrel created from a scalar, a character string, a vector, and a function:

```
> lst <- list(3.14, "Moe", c(1,1,2,3), mean)
> lst
[[1]]
[1] 3.14

[[2]]
```

```
[1] "Moe"

[[3]]
[1] 1 1 2 3

[[4]]
function (x, ...)
UseMethod("mean")
<environment: namespace:base>
```

You can also build a list by creating an empty list and populating it. Here is our "mongrel" example built in that way:

```
> lst <- list()
> lst[[1]] <- 3.14
> lst[[2]] <- "Moe"
> lst[[3]] <- c(1,1,2,3)
> lst[[4]] <- mean
```

List elements can be named. The `list` function lets you supply a name for every element:

```
> lst <- list(mid=0.5, right=0.841, far.right=0.977)
> lst
$mid
[1] 0.5

$right
[1] 0.841

$far.right
[1] 0.977
```

See Also

See the "Introduction" to this chapter for more about lists; see Recipe 5.9 for more about building and using lists with named elements.

5.7 Selecting List Elements by Position

Problem

You want to access list elements by position.

Solution

Use one of these ways. Here, `lst` is a list variable:

`lst[[n]]`
> Select the nth element from the list.

`lst[c(n_1, n_2, ..., n_k)]`
> Returns a list of elements, selected by their positions.

Note that the first form returns a single element and the second returns a list.

Discussion

Suppose we have a list of four integers, called years:

```
> years <- list(1960, 1964, 1976, 1994)
> years
[[1]]
[1] 1960

[[2]]
[1] 1964

[[3]]
[1] 1976

[[4]]
[1] 1994
```

We can access single elements using the double-square-bracket syntax:

```
> years[[1]]
[1] 1960
```

We can extract sublists using the single-square-bracket syntax:

```
> years[c(1,2)]
[[1]]
[1] 1960

[[2]]
[1] 1964
```

This syntax can be confusing because of a subtlety: there is an important difference between lst[[*n*]] and lst[*n*]. They are not the same thing:

lst[[*n*]]
> This is an element, not a list. It is the *n*th element of lst.

lst[*n*]
> This is a list, not an element. The list contains one element, taken from the *n*th element of lst. This is a special case of lst[c(n_1, n_2, ..., n_k)] in which we eliminated the c(...) construct because there is only one n.

The difference becomes apparent when we inspect the structure of the result—one is a number; the other is a list:

```
> class(years[[1]])
[1] "numeric"

> class(years[1])
[1] "list"
```

The difference becomes annoyingly apparent when we cat the value. Recall that cat can print atomic values or vectors but complains about printing structured objects:

```
> cat(years[[1]], "\n")
1960
> cat(years[1], "\n")
Error in cat(list(...), file, sep, fill, labels, append) :
  argument 1 (type 'list') cannot be handled by 'cat'
```

We got lucky here because R alerted us to the problem. In other contexts, you might work long and hard to figure out that you accessed a sublist when you wanted an element, or vice versa.

5.8 Selecting List Elements by Name

Problem

You want to access list elements by their names.

Solution

Use one of these forms. Here, lst is a list variable:

lst[["name"]]
 Selects the element called *name*. Returns NULL if no element has that name.

lst$name
 Same as previous, just different syntax.

lst[c(name₁, name₂, ..., nameₖ)]
 Returns a list built from the indicated elements of lst.

Note that the first two forms return an element whereas the third form returns a list.

Discussion

Each element of a list can have a name. If named, the element can be selected by its name. This assignment creates a list of four named integers:

```
> years <- list(Kennedy=1960, Johnson=1964, Carter=1976, Clinton=1994)
```

These next two expressions return the same value—namely, the element that is named "Kennedy":

```
> years[["Kennedy"]]
[1] 1960
> years$Kennedy
[1] 1960
```

The following two expressions return sublists extracted from years:

```
> years[c("Kennedy","Johnson")]
$Kennedy
```

```
[1] 1960

$Johnson
[1] 1964

> years["Carter"]
$Carter
[1] 1976
```

Just as with selecting list elements by position (Recipe 5.7), there is an important difference between lst[["*name*"]] and lst["*name*"]. They are not the same:

lst[["*name*"]]
 This is an element, not a list.

lst["*name*"]
 This is a list, not an element. This is a special case of lst[c(*name₁, name₂, ...,* *name$_k$*)] in which we don't need the c(...) construct because there is only one name.

See Also

See Recipe 5.7 to access elements by position rather than by name.

5.9 Building a Name/Value Association List

Problem

You want to create a list that associates names and values—as would a dictionary, hash, or lookup table in another programming language.

Solution

The list function lets you give names to elements, creating an association between each name and its value:

```
> lst <- list(mid=0.5, right=0.841, far.right=0.977)
```

If you have parallel vectors of names and values, you can create an empty list and then populate the list by using a vectorized assignment statement:

```
> lst <- list()
> lst[names] <- values
```

Discussion

Each element of a list can be named, and you can retrieve list elements by name. This gives you a basic programming tool: the ability to associate names with values.

You can assign element names when you build the list. The list function allows arguments of the form *name=value*:

```
> lst <- list(
+           far.left=0.023,
+           left=0.159,
+           mid=0.500,
+           right=0.841,
+           far.right=0.977)
> lst
$far.left
[1] 0.023

$left
[1] 0.159

$mid
[1] 0.5

$right
[1] 0.841

$far.right
[1] 0.977
```

One way to name the elements is to create an empty list and then populate it via assignment statements:

```
> lst <- list()
> lst$far.left <- 0.023
> lst$left <- 0.159
> lst$mid <- 0.500
> lst$right <- 0.841
> lst$far.right <- 0.977
```

Sometimes you have a vector of names and a vector of corresponding values:

```
> values <- pnorm(-2:2)
> names <- c("far.left", "left", "mid", "right", "far.right")
```

You can associate the names and the values by creating an empty list and then populating it with a vectorized assignment statement:

```
> lst <- list()
> lst[names] <- values
> lst
$far.left
[1] 0.02275013

$left
[1] 0.1586553

$mid
[1] 0.5

$right
[1] 0.8413447

$far.right
[1] 0.9772499
```

Once the association is made, the list can "translate" names into values through a simple list lookup:

```
> cat("The left limit is", lst[["left"]], "\n")
The left limit is 0.1586553
> cat("The right limit is", lst[["right"]], "\n")
The right limit is 0.8413447

> for (nm in names(lst)) cat("The", nm, "limit is", lst[[nm]], "\n")
The far.left limit is 0.02275013
The left limit is 0.1586553
The mid limit is 0.5
The right limit is 0.8413447
The far.right limit is 0.9772499
```

5.10 Removing an Element from a List

Problem

You want to remove an element from a list.

Solution

Assign NULL to the element. R will remove it from the list.

Discussion

To remove a list element, select it by position or by name, and then assign NULL to the selected element:

```
> years
$Kennedy
[1] 1960

$Johnson
[1] 1964

$Carter
[1] 1976

$Clinton
[1] 1994

> years[["Johnson"]] <- NULL              # Remove the element labeled "Johnson"
> years
$Kennedy
[1] 1960

$Carter
[1] 1976

$Clinton
[1] 1994
```

You can remove multiple elements this way, too:

```
> years[c("Carter","Clinton")] <- NULL      # Remove two elements
> years
$Kennedy
[1] 1960
```

5.11 Flatten a List into a Vector

Problem

You want to flatten all the elements of a list into a vector.

Solution

Use the unlist function.

Discussion

There are many contexts that require a vector. Basic statistical functions work on vectors but not on lists, for example. If iq.scores is a list of numbers, then we cannot directly compute their mean:

```
> mean(iq.scores)
[1] NA
Warning message:
In mean.default(iq.scores) :
  argument is not numeric or logical: returning NA
```

Instead, we must flatten the list into a vector using unlist and then compute the mean of the result:

```
> mean(unlist(iq.scores))
[1] 106.4452
```

Here is another example. We can cat scalars and vectors, but we cannot cat a list:

```
> cat(iq.scores, "\n")
Error in cat(list(...), file, sep, fill, labels, append) :
  argument 1 (type 'list') cannot be handled by 'cat'
```

One solution is to flatten the list into a vector before printing:

```
> cat("IQ Scores:", unlist(iq.scores), "\n")
IQ Scores: 89.73383 116.5565 113.0454
```

See Also

Conversions such as this are discussed more fully in Recipe 5.33.

5.12 Removing NULL Elements from a List

Problem

Your list contains NULL values. You want to remove them.

Solution

Suppose lst is a list some of whose elements are NULL. This expression will remove the NULL elements:

```
> lst[sapply(lst, is.null)] <- NULL
```

Discussion

Finding and removing NULL elements from a list is surprisingly tricky. I wrote the following expression after trying several other ways, including the obvious ones, and failing. Here's how it works:

1. R calls sapply to apply the is.null function to every element of the list.
2. sapply returns a vector of logical values that are TRUE wherever the corresponding list element is NULL.
3. R selects values from the list according to that vector.
4. R assigns NULL to the selected items, removing them from the list.

The curious reader may be wondering how a list can contain NULL elements, given that we remove elements by setting them to NULL (Recipe 5.10). The answer is that we can create a list containing NULL elements:

```
> lst <- list("Moe", NULL, "Curly")     # Create list with NULL element
> lst
[[1]]
[1] "Moe"

[[2]]
NULL

[[3]]
[1] "Curly"

> lst[sapply(lst, is.null)] <- NULL     # Remove NULL element from list
> lst
[[1]]
[1] "Moe"

[[2]]
[1] "Curly"
```

See Also

See Recipe 5.10 for how to remove list elements.

5.13 Removing List Elements Using a Condition

Problem

You want to remove elements from a list according to a conditional test, such as removing elements that are negative or smaller than some threshold.

Solution

Build a logical vector based on the condition. Use the vector to select list elements and then assign NULL to those elements. This assignment, for example, removes all negative value from lst:

```
> lst[lst < 0] <- NULL
```

Discussion

This recipe is based on two useful features of R. First, a list can be indexed by a logical vector. Wherever the vector element is TRUE, the corresponding list element is selected. Second, you can remove a list element by assigning NULL to it.

Suppose we want to remove elements from lst whose value is zero. We construct a logical vector which identifies the unwanted values (lst == 0). Then we select those elements from the list and assign NULL to them:

```
> lst[lst == 0] <- NULL
```

This expression will remove NA values from the list:

```
> lst[is.na(lst)] <- NULL
```

So far, so good. The problems arise when you cannot easily build the logical vector. That often happens when you want to use a function that cannot handle a list. Suppose you want to remove list elements whose absolute value is less than 1. The abs function will not handle a list, unfortunately:

```
> lst[abs(lst) < 1] <- NULL
Error in abs(lst) : non-numeric argument to function
```

The simplest solution is flattening the list into a vector by calling unlist and then testing the vector:

```
> lst[abs(unlist(lst)) < 1] <- NULL
```

A more elegant solution uses lapply to apply the function to every element of the list:

```
> lst[lapply(lst,abs) < 1] <- NULL
```

Lists can hold complex objects, too, not just atomic values. Suppose that mods is a list of linear models created by the lm function. This expression will remove any model whose R^2 value is less than 0.30:

```
> mods[sapply(mods, function(m) summary(m)$r.squared < 0.3)] <- NULL
```

See Also

See Recipes 5.7, 5.10, 5.11, 6.2, and 11.1.

5.14 Initializing a Matrix

Problem

You want to create a matrix and initialize it from given values.

Solution

Capture the data in a vector or list, and then use the matrix function to shape the data into a matrix. This example shapes a vector into a 2 × 3 matrix (i.e., two rows and three columns):

```
> matrix(vec, 2, 3)
```

Discussion

Suppose we want to create and initialize a 2 × 3 matrix. We can capture the initial data inside a vector and then shape it using the matrix function:

```
> theData <- c(1.1, 1.2, 2.1, 2.2, 3.1, 3.2)
> mat <- matrix(theData, 2, 3)
> mat
     [,1] [,2] [,3]
[1,]  1.1  2.1  3.1
[2,]  1.2  2.2  3.2
```

The first argument of matrix is the data, the second argument is the number of rows, and the third argument is the number of columns. Observe that the matrix was filled column by column, not row by row.

It's common to initialize an entire matrix to one value such as zero or NA. If the first argument of matrix is a single value, then R will apply the Recycling Rule and automatically replicate the value to fill the entire matrix:

```
> matrix(0, 2, 3)          # Create an all-zeros matrix
     [,1] [,2] [,3]
[1,]   0    0    0
[2,]   0    0    0
> matrix(NA, 2, 3)         # Create a matrix populated with NA
     [,1] [,2] [,3]
```

```
[1,]   NA   NA   NA
[2,]   NA   NA   NA
```

You can create a matrix with a one-liner, of course, but it becomes difficult to read:

```
> mat <- matrix(c(1.1, 1.2, 1.3, 2.1, 2.2, 2.3), 2, 3)
```

A common idiom in R is typing the data itself in a rectangular shape that reveals the matrix structure:

```
> theData <- c(1.1, 1.2, 1.3,
+              2.1, 2.2, 2.3)
> mat <- matrix(theData, 2, 3, byrow=TRUE)
```

Setting byrow=TRUE tells matrix that the data is row-by-row and not column-by-column (which is the default). In condensed form, that becomes:

```
> mat <- matrix(c(1.1, 1.2, 1.3,
+                 2.1, 2.2, 2.3),
+               2, 3, byrow=TRUE)
```

Expressed this way, the reader quickly sees the two rows and three columns of data.

There is a quick-and-dirty way to turn a vector into a matrix: just assign dimensions to the vector. This was discussed in the "Introduction". The following example creates a vanilla vector and then shapes it into a 2 × 3 matrix:

```
> v <- c(1.1, 1.2, 1.3, 2.1, 2.2, 2.3)
> dim(v) <- c(2,3)
> v
     [,1] [,2] [,3]
[1,]  1.1  1.3  2.2
[2,]  1.2  2.1  2.3
```

Personally, I find this more opaque than using matrix, especially since there is no byrow option here.

See Also

See Recipe 5.3.

5.15 Performing Matrix Operations

Problem

You want to perform matrix operations such as transpose, matrix inversion, matrix multiplication, or constructing an identity matrix.

Solution

t(A)

Matrix transposition of A

```
solve(A)
```
 Matrix inverse of A
```
A %*% B
```
 Matrix multiplication of A and B
```
diag(n)
```
 An *n*-by-*n* diagonal (identity) matrix

Discussion

Recall that A*B is element-wise multiplication whereas A %*% B is matrix multiplication.

All these functions return a matrix. Their arguments can be either matrices or data frames. If they are data frames then R will first convert them to matrices (although this is useful only if the data frame contains exclusively numeric values).

5.16 Giving Descriptive Names to the Rows and Columns of a Matrix

Problem

You want to assign descriptive names to the rows or columns of a matrix.

Solution

Every matrix has a rownames attribute and a colnames attribute. Assign a vector of character strings to the appropriate attribute:

```
> rownames(mat) <- c("rowname₁", "rowname₂", ..., "rownameₘ")
> colnames(mat) <- c("colname₁", "colname₂", ..., "colnameₙ")
```

Discussion

R lets you assign names to the rows and columns of a matrix, which is useful for printing the matrix. R will display the names if they are defined, enhancing the readability of your output. Consider this matrix of correlations between the prices of IBM, Microsoft, and Google stock:

```
> print(tech.corr)
       [,1]  [,2]  [,3]
[1,] 1.000 0.556 0.390
[2,] 0.556 1.000 0.444
[3,] 0.390 0.444 1.000
```

In this form, the matrix output's interpretation is not self-evident. Yet if we define names for the rows and columns, then R will annotate the matrix output with the names:

```
> colnames(tech.corr) <- c("IBM","MSFT","GOOG")
> rownames(tech.corr) <- c("IBM","MSFT","GOOG")
> print(tech.corr)
       IBM  MSFT  GOOG
IBM  1.000 0.556 0.390
MSFT 0.556 1.000 0.444
GOOG 0.390 0.444 1.000
```

Now the reader knows at a glance which rows and columns apply to which stocks.

Another advantage of naming rows and columns is that you can refer to matrix elements by those names:

```
> tech.corr["IBM","GOOG"]     # What is the correlation between IBM and GOOG?
[1] 0.39
```

5.17 Selecting One Row or Column from a Matrix

Problem

You want to select a single row or a single column from a matrix.

Solution

The solution depends on what you want. If you want the result to be a simple vector, just use normal indexing:

```
> vec <- mat[1,]       # First row
> vec <- mat[,3]       # Third column
```

If you want the result to be a one-row matrix or a one-column matrix, then include the drop=FALSE argument:

```
> row <- mat[1,,drop=FALSE]     # First row in a one-row matrix
> col <- mat[,3,drop=FALSE]     # Third column in a one-column matrix
```

Discussion

Normally, when you select one row or column from a matrix, R strips off the dimensions. The result is a dimensionless vector:

```
> mat[1,]
[1]  1  4  7 10
> mat[,3]
[1] 7 8 9
```

When you include the drop=FALSE argument, however, R retains the dimensions. In that case, selecting a row returns a row vector (a $1 \times n$ matrix):

```
> mat[1,,drop=FALSE]
     [,1] [,2] [,3] [,4]
[1,]    1    4    7   10
```

Likewise, selecting a column with drop=FALSE returns a column vector (an $n \times 1$ matrix):

```
> mat[,3,drop=FALSE]
     [,1]
[1,]    7
[2,]    8
[3,]    9
```

5.18 Initializing a Data Frame from Column Data

Problem

Your data is organized by columns, and you want to assemble it into a data frame.

Solution

If your data is captured in several vectors and/or factors, use the data.frame function to assemble them into a data frame:

```
> dfrm <- data.frame(v1, v2, v3, f1, f2)
```

If your data is captured in a *list* that contains vectors and/or factors, use instead as.data.frame:

```
> dfrm <- as.data.frame(list.of.vectors)
```

Discussion

A data frame is a collection of columns, each of which corresponds to an observed variable (in the statistical sense, not the programming sense). If your data is already organized into columns, then it's easy to build a data frame.

The data.frame function can construct a data frame from vectors, where each vector is one observed variable. Suppose you have two numeric predictor variables, one categorical predictor variable, and one response variable. The data.frame function can create a data frame from your vectors:

```
> dfrm <- data.frame(pred1, pred2, pred3, resp)
> dfrm
        pred1        pred2 pred3     resp
1  -2.7528917 -1.40784130    AM 12.57715
2  -0.3626909  0.31286963    AM 21.02418
3  -1.0416039 -0.69685664    PM 18.94694
4   1.2666820 -1.27511434    PM 18.98153
5   0.7806372 -0.27292745    AM 19.59455
6  -1.0832624  0.73383339    AM 20.71605
7  -2.0883305  0.96816822    PM 22.70062
8  -0.7063653 -0.84476203    PM 18.40691
9  -0.8394022  0.31530793    PM 21.00930
10 -0.4966884 -0.08030948    AM 19.31253
```

Notice that `data.frame` takes the column names from your program variables. You can override that default by supplying explicit column names:

```
> dfrm <- data.frame(p1=pred1, p2=pred2, p3=pred3, r=resp)
> dfrm
          p1         p2 p3       r
1  -2.7528917 -1.40784130 AM 12.57715
2  -0.3626909  0.31286963 AM 21.02418
3  -1.0416039 -0.69685664 PM 18.94694
.
. (etc.)
.
```

Alternatively, your data may be organized into vectors but those vectors are held in a list, not individual program variables, like this:

```
> lst <- list(p1=pred1, p2=pred2, p3=pred3, r=resp)
```

No problem. Use the `as.data.frame` function to create a data frame from the list of vectors:

```
> as.data.frame(lst)
          p1         p2 p3       r
1  -2.7528917 -1.40784130 AM 12.57715
2  -0.3626909  0.31286963 AM 21.02418
3  -1.0416039 -0.69685664 PM 18.94694
.
. (etc.)
.
```

5.19 Initializing a Data Frame from Row Data

Problem

Your data is organized by rows, and you want to assemble it into a data frame.

Solution

Store each row in a one-row data frame. Store the one-row data frames in a list. Use `rbind` and `do.call` to bind the rows into one, large data frame:

```
> dfrm <- do.call(rbind, obs)
```

Here, `obs` is a list of one-row data frames.

Discussion

Data often arrives as a collection of observations. Each *observation* is a record or tuple that contains several values, one for each observed variable. The lines of a flat file are usually like that: each line is one record, each record contains several columns, and each column is a different variable (see Recipe 4.12). Such data is organized by

observation, not by *variable*. In other words, you are given rows one at a time rather than columns one at a time.

Each such row might be stored in several ways. One obvious way is as a vector. If you have purely numerical data, use a vector.

However, many datasets are a mixture of numeric, character, and categorical data, in which case a vector won't work. I recommend storing each such heterogeneous row in a one-row data frame. (You could store each row in a list, but this recipe gets a little more complicated.)

For concreteness, let's assume that you have ten rows with four variables per observation: `pred1`, `pred2`, `pred2`, and `resp`. Each row is stored in a one-row data frame, so you have ten such data frames. Those data frames are stored in a list called `obs`. The first element of `obs` might look like this:

```
> obs[[1]]
   pred1 pred2 pred3   resp
1 -1.197  0.36    AM 18.701
```

This recipe works also if your observations are stored in vectors rather than one-row data frames.

We need to bind together those rows into a data frame. That's what the `rbind` function does. It binds its arguments in such a way that each argument becomes one row in the result. If we `rbind` the first two observations, for example, we get a two-row data frame:

```
> rbind(obs[[1]], obs[[2]])
   pred1 pred2 pred3   resp
1 -1.197  0.36    AM 18.701
2 -0.952  1.23    PM 25.709
```

We want to bind together every observation, not just the first two, so we tap into the vector processing of R. The `do.call` function will expand `obs` into one, long argument list and call `rbind` with that long argument list:

```
> do.call(rbind,obs)
    pred1  pred2 pred3   resp
1  -1.197  0.360    AM 18.701
2  -0.952  1.230    PM 25.709
3   0.279  0.423    PM 21.572
4  -1.445 -1.846    AM 14.392
5   0.822 -0.246    AM 19.841
6   1.247  1.254    PM 25.637
7  -0.394  1.563    AM 24.585
8  -1.248 -1.264    PM 16.770
9  -0.652 -2.344    PM 14.915
10 -1.171 -0.776    PM 17.948
```

The result is a data frame built from our rows of data.

Sometimes, for reasons beyond your control, the rows of your data are stored in lists rather than one-row data frames. You may be dealing with rows returned by a database package, for example. In that case, `obs` will be a list of lists, not a list of data frames.

We first transform the rows into data frames using the Map function and then apply this recipe:

```
> dfrm <- do.call(rbind,Map(as.data.frame,obs))
```

See Also

See Recipe 5.18 if your data is organized by columns, not rows; see Recipe 12.18 to learn more about do.call.

5.20 Appending Rows to a Data Frame

Problem

You want to append one or more new rows to a data frame.

Solution

Create a second, temporary data frame containing the new rows. Then use the rbind function to append the temporary data frame to the original data frame.

Discussion

Suppose we want to append a new row to our data frame of Chicago-area cities. First, we create a one-row data frame with the new data:

```
> newRow <- data.frame(city="West Dundee", county="Kane", state="IL", pop=5428)
```

Next, we use the rbind function to append that one-row data frame to our existing data frame:

```
> suburbs <- rbind(suburbs, newRow)
> suburbs
                  city  county state     pop
1              Chicago    Cook    IL 2853114
2              Kenosha Kenosha    WI   90352
3               Aurora    Kane    IL  171782
4                Elgin    Kane    IL   94487
5                 Gary Lake(IN)    IN  102746
6               Joliet Kendall    IL  106221
7           Naperville  DuPage    IL  147779
8    Arlington Heights    Cook    IL   76031
9          Bolingbrook    Will    IL   70834
10              Cicero    Cook    IL   72616
11            Evanston    Cook    IL   74239
12             Hammond Lake(IN)    IN   83048
13            Palatine    Cook    IL   67232
14          Schaumburg    Cook    IL   75386
15              Skokie    Cook    IL   63348
16            Waukegan Lake(IL)    IL   91452
17         West Dundee    Kane    IL    5428
```

The `rbind` function tells R that we are appending a new row to `suburbs`, not a new column. It may be obvious to you that `newRow` is a row and not a column, but it is not obvious to R. (Use the `cbind` function to append a column.)

One word of caution. The new row must use the same column names as the data frame. Otherwise, `rbind` will fail.

We can combine these two steps into one, of course:

```
> suburbs <- rbind(suburbs,
+               data.frame(city="West Dundee", county="Kane", state="IL", pop=5428))
```

We can even extend this technique to multiple new rows because `rbind` allows multiple arguments:

```
> suburbs <- rbind(suburbs,
+               data.frame(city="West Dundee", county="Kane", state="IL", pop=5428),
+               data.frame(city="East Dundee", county="Kane", state="IL", pop=2955))
```

 Do not use this recipe to append many rows to a large data frame. That would force R to reallocate a large data structure repeatedly, which is a very slow process. Build your data frame using more efficient means, such as those in Recipes 5.19 or 5.21.

5.21 Preallocating a Data Frame

Problem

You are building a data frame, row by row. You want to preallocate the space instead of appending rows incrementally.

Solution

Create a data frame from generic vectors and factors using the functions `numeric(n)`, `character(n)`, and `factor(n)`:

```
> dfrm <- data.frame(colname1=numeric(n), colname2=character(n), ... etc. ... )
```

Here, *n* is the number of rows needed for the data frame.

Discussion

Theoretically, you can build a data frame by appending new rows, one by one. That's OK for small data frames, but building a large data frame in that way can be tortuous. The memory manager in R works poorly when one new row is repeatedly appended to a large data structure. Hence your R code will run very slowly.

One solution is to preallocate the data frame—assuming you know the required number of rows. By preallocating the data frame once and for all, you sidestep problems with the memory manager.

Suppose you want to create a data frame with 1,000,000 rows and three columns: two numeric and one character. Use the `numeric` and `character` functions to preallocate the columns; then join them together using `data.frame`:

```
> N <- 1000000
> dfrm <- data.frame(dosage=numeric(N), lab=character(N), response=numeric(N))
```

Now you have a data frame with the correct dimensions, 1,000,000 × 3, waiting to receive its contents.

Data frames can contain factors, but preallocating a factor is a little trickier. You can't simply call `factor(n)`. You need to specify the factor's levels because you are creating it. Continuing our example, suppose you want the `lab` column to be a factor, not a character string, and that the possible levels are NJ, IL, and CA. Include the levels in the column specification, like this:

```
> N <- 1000000
> dfrm <- data.frame(dosage=numeric(N),
+                    lab=factor(N, levels=c("NJ", "IL", "CA")),
+                    response=numeric(N) )
```

5.22 Selecting Data Frame Columns by Position

Problem

You want to select columns from a data frame according to their position.

Solution

To select a single column, use this list operator:

`dfrm[[n]]`
> Returns *one column*—specifically, the *n*th column of `dfrm`.

To select one or more columns and package them in a data frame, use the following sublist expressions:

`dfrm[n]`
> Returns a *data frame* consisting solely of the *n*th column of `dfrm`.

`dfrm[c(n_1, n_2, ..., n_k)]`
> Returns a *data frame* built from the columns in positions $n_1, n_2, ..., n_k$ of `dfrm`.

You can use matrix-style subscripting to select one or more columns:

`dfrm[, n]`
> Returns the *n*th column (assuming that *n* contains exactly one value).

`dfrm[, c(n_1, n_2, ..., n_k)]`
> Returns a *data frame* built from the columns in positions $n_1, n_2, ..., n_k$.

Note that the matrix-style subscripting can return two different data types (either column or data frame) depending upon whether you select one column or multiple columns.

Discussion

There are a bewildering number of ways to select columns from a data frame. The choices can be confusing until you understand the logic behind the alternatives. As you read this explanation, notice how a slight change in syntax—a comma here, a double-bracket there—changes the meaning of the expression.

Let's play with the population data for the 16 largest cities in the Chicago metropolitan area:

```
> suburbs
                 city   county state      pop
1             Chicago     Cook    IL  2853114
2             Kenosha  Kenosha    WI    90352
3              Aurora     Kane    IL   171782
4               Elgin     Kane    IL    94487
5                Gary Lake(IN)    IN   102746
6              Joliet  Kendall    IL   106221
7          Naperville   DuPage    IL   147779
8   Arlington Heights     Cook    IL    76031
9         Bolingbrook     Will    IL    70834
10             Cicero     Cook    IL    72616
11           Evanston     Cook    IL    74239
12            Hammond Lake(IN)    IN    83048
13           Palatine     Cook    IL    67232
14         Schaumburg     Cook    IL    75386
15             Skokie     Cook    IL    63348
16           Waukegan Lake(IL)    IL    91452
```

Use simple list notation to select exactly one column, such as the first column:

```
> suburbs[[1]]
 [1] "Chicago"      "Kenosha"      "Aurora"       "Elgin"
 [5] "Gary"         "Joliet"       "Naperville"   "Arlington Heights"
 [9] "Bolingbrook"  "Cicero"       "Evanston"     "Hammond"
[13] "Palatine"     "Schaumburg"   "Skokie"       "Waukegan"
```

The first column of suburbs is a vector, so that's what suburbs[[1]] returns: a vector. If the first column were a factor, we'd get a factor.

The result differs when you use the single-bracket notation, as in suburbs[1] or suburbs[c(1,3)]. You still get the requested columns, but R wraps them in a data frame. This example returns the first column wrapped in a data frame:

```
> suburbs[1]
         city
1     Chicago
2     Kenosha
3      Aurora
4       Elgin
```

```
 5              Gary
 6            Joliet
 7        Naperville
 8  Arlington Heights
 9       Bolingbrook
10            Cicero
11          Evanston
12           Hammond
13          Palatine
14        Schaumburg
15            Skokie
16          Waukegan
```

The next example returns the first and third columns wrapped in a data frame:

```
> suburbs[c(1,3)]
                city     pop
1            Chicago 2853114
2            Kenosha   90352
3             Aurora  171782
4              Elgin   94487
5               Gary  102746
6             Joliet  106221
7         Naperville  147779
8  Arlington Heights   76031
9        Bolingbrook   70834
10            Cicero   72616
11          Evanston   74239
12           Hammond   83048
13          Palatine   67232
14        Schaumburg   75386
15            Skokie   63348
16          Waukegan   91452
```

A major source of confusion is that suburbs[[1]] and suburbs[1] look similar but produce very different results:

suburbs[[1]]

This returns one column.

suburbs[1]

This returns a data frame, and the data frame contains exactly one column. This is a special case of dfrm[c(n_1,n_2, ..., n_k)]. We don't need the c(...) construct because there is only one n.

The point here is that "one column" is different from "a data frame that contains one column." The first expression returns a column, so it's a vector or a factor. The second expression returns a data frame, which is different.

R lets you use matrix notation to select columns, as shown in the Solution. But an odd quirk can bite you: you might get a column or you might get a data frame, depending upon many subscripts you use. In the simple case of one index you get a column, like this:

```
> suburbs[,1]
 [1] "Chicago"      "Kenosha"     "Aurora"      "Elgin"
 [5] "Gary"         "Joliet"      "Naperville"  "Arlington Heights"
 [9] "Bolingbrook"  "Cicero"      "Evanston"    "Hammond"
[13] "Palatine"     "Schaumburg"  "Skokie"      "Waukegan"
```

But using the same matrix-style syntax with multiple indexes returns a data frame:

```
> suburbs[,c(1,4)]
                city      pop
1            Chicago  2853114
2            Kenosha    90352
3             Aurora   171782
4              Elgin    94487
5               Gary   102746
6             Joliet   106221
7         Naperville   147779
8  Arlington Heights    76031
9        Bolingbrook    70834
10            Cicero    72616
11          Evanston    74239
12           Hammond    83048
13          Palatine    67232
14         Schaumburg   75386
15            Skokie    63348
16          Waukegan    91452
```

This creates a problem. Suppose you see this expression in some old R script:

```
dfrm[,vec]
```

Quick, does that return a column or a data frame? Well, it depends. If vec contains one value then you get a column; otherwise, you get a data frame. You cannot tell from the syntax alone.

To avoid this problem, you can include drop=FALSE in the subscripts; this forces R to return a data frame:

```
dfrm[,vec,drop=FALSE]
```

Now there is no ambiguity about the returned data structure. It's a data frame.

When all is said and done, using matrix notation to select columns from data frames is not the best procedure. I recommend that you instead use the list operators described previously. They just seem clearer.

See Also

See Recipe 5.17 for more about using drop=FALSE.

5.23 Selecting Data Frame Columns by Name

Problem

You want to select columns from a data frame according to their name.

Solution

To select a single column, use one of these list expressions:

dfrm[["*name*"]]
> Returns *one column*, the column called *name*.

dfrm$*name*
> Same as previous, just different syntax.

To select one or more columns and package them in a data frame, use these list expressions:

dfrm["*name*"]
> Selects one column and packages it inside a data frame object.

dfrm[c("*name₁*", "*name₂*", ..., "*nameₖ*")]
> Selects several columns and packages them in a data frame.

You can use matrix-style subscripting to select one or more columns:

dfrm[, "*name*"]
> Returns the named column.

dfrm[, c("*name₁*", "*name₂*", ..., "*nameₖ*")]
> Selects several columns and packages in a data frame.

Once again, the matrix-style subscripting can return two different data types (column or data frame) depending upon whether you select one column or multiple columns.

Discussion

All columns in a data frame must have names. If you know the name, it's usually more convenient and readable to select by name, not by position.

The solutions just described are similar to those for Recipe 5.22, where we selected columns by position. The only difference is that here we use column names instead of column numbers. All the observations made in Recipe 5.22 apply here:

- dfrm[["name"]] returns one column, not a data frame.
- dfrm[c("*name₁*", "*name₂*", ..., "*nameₖ*")] returns a data frame, not a column.
- dfrm["name"] is a special case of the previous expression and so returns a data frame, not a column.

- The matrix-style subscripting can return either a column or a data frame, so be careful how many names you supply. See Recipe 5.22 for a discussion of this "gotcha" and using drop=FALSE.

There is one new addition:

```
dfrm$name
```

This is identical in effect to dfrm[["name"]], but it's easier to type and to read.

See Also

See Recipe 5.22 to understand these ways to select columns.

5.24 Selecting Rows and Columns More Easily

Problem

You want an easier way to select rows and columns from a data frame or matrix.

Solution

Use the subset function. The select argument is a column name, or a vector of column names, to be selected:

```
> subset(dfrm, select=colname)
> subset(dfrm, select=c(colname₁, ..., colnameₙ))
```

Note that you do *not* quote the column names.

The subset argument is a logical expression that selects rows. Inside the expression, you can refer to the column names as part of the logical expression. In this example, response is a column in the data frame, and we are selecting rows with a positive response:

```
> subset(dfrm, subset=(response > 0))
```

subset is most useful when you combine the select and subset arguments:

```
> subset(dfrm, select=c(predictor,response), subset=(response > 0))
```

Discussion

Indexing is the "official" way to select rows and columns from a data frame, as described in Recipes 5.22 and 5.23. However, indexing is cumbersome when the index expressions become complicated.

The subset function provides a more convenient and readable way to select rows and columns. It's beauty is that you can refer to the columns of the data frame right inside the expressions for selecting columns and rows.

Here are some examples using the `Cars93` dataset in the `MASS` package. Recall that the dataset includes columns for `Manufacturer`, `Model`, `MPG.city`, `MPG.highway`, `Min.Price`, and `Max.Price`:

Select the model name for cars that can exceed 30 miles per gallon (MPG) in the city

```
> subset(Cars93, select=Model, subset=(MPG.city > 30))
      Model
31 Festiva
39   Metro
42   Civic
.
. (etc.)
.
```

Select the model name and price range for four-cylinder cars made in the United States

```
> subset(Cars93, select=c(Model,Min.Price,Max.Price),
+       subset=(Cylinders == 4 & Origin == "USA"))
        Model Min.Price Max.Price
6     Century      14.2      17.3
12   Cavalier       8.5      18.3
13    Corsica      11.4      11.4
.
. (etc.)
.
```

Select the manufacturer's name and the model name for all cars whose highway MPG value is above the median

```
> subset(Cars93, select=c(Manufacturer,Model),
+       subset=c(MPG.highway > median(MPG.highway)))
   Manufacturer    Model
1         Acura  Integra
5           BMW     535i
6         Buick  Century
.
. (etc.)
.
```

The `subset` function is actually more powerful than this recipe implies. It can select from lists and vectors, too. See the help page for details.

5.25 Changing the Names of Data Frame Columns

Problem

You converted a matrix or list into a data frame. R gave names to the columns, but the names are at best uninformative and at worst bizarre.

Solution

Data frames have a `colnames` attribute that is a vector of column names. You can update individual names or the entire vector:

```
> colnames(dfrm) <- newnames        # newnames is a vector of character strings
```

Discussion

The columns of data frames must have names. If you convert a vanilla matrix into a data frame, R will synthesize names that are reasonable but boring—for example, V1, V2, V3, and so forth:

```
> mat
       [,1]   [,2]   [,3]
[1,] -0.818 -0.667 -0.494
[2,] -0.819 -0.946 -0.205
[3,]  0.385  1.531 -0.611
[4,] -2.155 -0.535 -0.316
> as.data.frame(mat)
      V1     V2     V3
1 -0.818 -0.667 -0.494
2 -0.819 -0.946 -0.205
3  0.385  1.531 -0.611
4 -2.155 -0.535 -0.316
```

If the matrix had column names defined, R would have used those names instead of synthesizing new ones.

However, converting a list into a data frame produces some strange synthetic names:

```
> lst
[[1]]
[1] -0.284 -1.114 -1.097 -0.873

[[2]]
[1] -1.673  0.929  0.306  0.778

[[3]]
[1]  0.323  0.368  0.067 -0.080

> as.data.frame(lst)
  c..0.284...1.114...1.097...0.873. c..1.673..0.929..0.306..0.778.
1                            -0.284                         -1.673
2                            -1.114                          0.929
3                            -1.097                          0.306
4                            -0.873                          0.778
  c.0.323..0.368..0.067...0.08.
1                         0.323
2                         0.368
3                         0.067
4                        -0.080
```

Again, if the list elements had names then R would have used them.

Fortunately, you can overwrite the synthetic names with names of your own by setting the `colnames` attribute:

```
> dfrm <- as.data.frame(lst)
> colnames(dfrm) <- c("before","treatment","after")
> dfrm
  before treatment  after
1 -0.284    -1.673  0.323
2 -1.114     0.929  0.368
3 -1.097     0.306  0.067
4 -0.873     0.778 -0.080
```

See Also

See Recipe 5.33.

5.26 Editing a Data Frame

Problem

Data in a data frame are incorrect or missing. You want a convenient way to edit the data frame contents.

Solution

R includes a data editor that displays your data frame in a spreadsheet-like window. Invoke the editor using the `edit` function:

```
> temp <- edit(dfrm)
> dfrm <- temp              # Overwrite only if you're happy with the changes!
```

Make your changes and then close the editor window. The edit function will return the updated data, which here are assigned to `temp`. If you are happy with the changes, overwrite your data frame with the results.

If you are feeling brave, the `fix` function invokes the editor and overwrites your variable with the result. There is no "undo", however:

```
> fix(dfrm)
```

Discussion

Figure 5-1 is a screenshot of the data editor on Windows during the editing of data from Recipe 5.22.

As of this writing, the editor is quite primitive. It does not include the common features of a modern editor—not even an "undo" command, for instance. I cannot recommend the data editor for regular use, but it's OK for emergency touch-ups.

Figure 5-1. Editing a data frame

Since there is no undo, take note of the solution that assigns the edited result to a temporary, intermediate variable. If you mess up your data, you can just delete the temporary variable without affecting the original data.

See Also

Several of the add-on GUI frontends provide data editors that are better than the native editor.

5.27 Removing NAs from a Data Frame

Problem

Your data frame contains NA values, which is creating problems for you.

Solution

Use `na.omit` to remove rows that contain any NA values.

```
> clean <- na.omit(dfrm)
```

Discussion

I frequently stumble upon situations where just a few NA values in my data frame cause everything to fall apart. One solution is simply to remove all rows that contain NAs. That's what na.omit does.

Here we can see cumsum fail because the input contains NA values:

```
> dfrm
          x          y
1 -0.9714511 -0.4578746
2         NA  3.1663282
3  0.3367627         NA
4  1.7520504  0.7406335
5  0.4918786  1.4543427
> cumsum(dfrm)
         x          y
1 -0.971451 -0.4578746
2        NA  2.7084536
3        NA         NA
4        NA         NA
5        NA         NA
```

If we remove the NA values, cumsum can complete its summations:

```
> cumsum(na.omit(dfrm))
          x          y
1 -0.9714511 -0.4578746
4  0.7805993  0.2827589
5  1.2724779  1.7371016
```

This recipe works for vectors and matrices, too, but not for lists.

Will You Still Have Enough Data?

The obvious danger here is that simply dropping observations from your data could render the results computationally or statistically meaningless. Make sure that omitting data makes sense in your context. Remember that na.omit will remove entire rows, not just the NA values, which could eliminate a lot of useful information.

5.28 Excluding Columns by Name

Problem

You want to exclude a column from a data frame using its name.

Solution

Use the subset function with a negated argument for the select parameter:

```
> subset(dfrm, select = -badboy)        # All columns except badboy
```

Discussion

We can exclude a column by position (e.g., dfrm[-1]), but how do we exclude a column by name? The subset function can exclude columns from a data frame. The select parameter is a normally a list of columns to include, but prefixing a minus sign (-) to the name causes the column to be excluded instead.

I often encounter this problem when calculating the correlation matrix of a data frame and I want to exclude nondata columns such as labels:

```
> cor(patient.data)
            patient.id         pre      dosage         post
patient.id  1.00000000  0.02286906   0.3643084  -0.13798149
pre         0.02286906  1.00000000   0.2270821  -0.03269263
dosage      0.36430837  0.22708208   1.0000000  -0.42006280
post       -0.13798149 -0.03269263  -0.4200628   1.00000000
```

This correlation matrix includes the meaningless "correlation" between patient ID and other variables, which is annoying. We can exclude the patient ID column to clean up the output:

```
> cor(subset(patient.data, select = -patient.id))
               pre      dosage         post
pre     1.00000000   0.2270821  -0.03269264
dosage  0.22708207   1.0000000  -0.42006280
post   -0.03269264  -0.4200628   1.00000000
```

We can exclude multiple columns by giving a vector of negated names:

```
> cor(subset(patient.data, select = c(-patient.id,-dosage)))
            pre         post
pre   1.00000000  -0.03269264
post -0.03269264   1.00000000
```

See Also

See Recipe 5.24 for more about the subset function.

5.29 Combining Two Data Frames

Problem

You want to combine the contents of two data frames into one data frame.

Solution

To combine the columns of two data frames side by side, use cbind:

```
> all.cols <- cbind(dfrm1,dfrm2)
```

To "stack" the rows of two data frames, use rbind:

```
> all.rows <- rbind(dfrm1,dfrm2)
```

Discussion

You can combine data frames in one of two ways: either by putting the columns side by side to create a wider data frame; or by "stacking" the rows to create a taller data frame. The cbind function will combine data frames side by side, as shown here when combining stooges and birth:

```
> stooges
   name n.marry n.child
1   Moe       1       2
2 Larry       1       2
3 Curly       4       2
> birth
  birth.year  birth.place
1       1887   Bensonhurst
2       1902  Philadelphia
3       1903      Brooklyn
> cbind(stooges,birth)
   name n.marry n.child birth.year  birth.place
1   Moe       1       2       1887   Bensonhurst
2 Larry       1       2       1902  Philadelphia
3 Curly       4       2       1903      Brooklyn
```

You would normally combine columns with the same height (number of rows). Technically speaking, however, cbind does not require matching heights. If one data frame is short, it will invoke the Recycling Rule to extend the short columns as necessary (Recipe 5.3), which may or may not be what you want.

The rbind function will "stack" the rows of two data frames, as shown here when combining stooges and guys:

```
> stooges
   name n.marry n.child
1   Moe       1       2
2 Larry       1       2
3 Curly       4       2
> guys
   name n.marry n.child
1   Tom       4       2
2  Dick       1       4
3 Harry       1       1
> rbind(stooges,guys)
   name n.marry n.child
1   Moe       1       2
2 Larry       1       2
3 Curly       4       2
4   Tom       4       2
5  Dick       1       4
6 Harry       1       1
```

The rbind function requires that the data frames have the same width: same number of columns and same column names. The columns need not be in the same *order*, however; rbind will sort that out.

Finally, this recipe is slightly more general than the title implies. First, you can combine more than two data frames because both `rbind` and `cbind` accept multiple arguments. Second, you can apply this recipe to other data types because `rbind` and `cbind` work also with vectors, lists, and matrices.

See Also

The `merge` function can combine data frames that are otherwise incompatible owing to missing or different columns. The `reshape2` and `plyr` packages, available on CRAN, include some powerful functions for slicing, dicing, and recombining data frames.

5.30 Merging Data Frames by Common Column

Problem

You have two data frames that share a common column. You want to merge their rows into one data frame by matching on the common column.

Solution

Use the `merge` function to join the data frames into one new data frame based on the common column:

```
> m <- merge(df1, df2, by="name")
```

Here *name* is the name of the column that is common to data frames *df1* and *df2*.

Discussion

Suppose you have two data frames, `born` and `died`, that each contain a column called name:

```
> born
    name year.born    place.born
1    Moe      1887   Bensonhurst
2  Larry      1902  Philadelphia
3  Curly      1903      Brooklyn
4  Harry      1964        Moscow
> died
    name year.died
1  Curly      1952
2    Moe      1975
3  Larry      1975
```

We can merge them into one data frame by using name to combine matched rows:

```
> merge(born, died, by="name")
    name year.born    place.born year.died
1  Curly      1903      Brooklyn      1952
2  Larry      1902  Philadelphia      1975
3    Moe      1887   Bensonhurst      1975
```

Notice that merge does not require the rows to be sorted or even to occur in the same order. It found the matching rows for Curly even though they occur in different positions. It also discards rows that appear in only one data frame or the other.

In SQL terms, the merge function essentially performs a join operation on the two data frames. It has many options for controlling that join operation, all of which are described on the help page for merge.

See Also

See Recipe 5.29 for other ways to combine data frames.

5.31 Accessing Data Frame Contents More Easily

Problem

Your data is stored in a data frame. You are getting tired of repeatedly typing the data frame name and want to access the columns more easily.

Solution

For quick, one-off expressions, use the with function to expose the column names:

```
> with(dataframe, expr)
```

Inside *expr*, you can refer to the columns of *dataframe* by their names—as if they were simple variables.

For repetitive access, use the attach function to insert the data frame into your search list. You can then refer to the data frame columns by name without mentioning the data frame:

```
> attach(dataframe)
```

Use the detach function to remove the data frame from your search list.

Discussion

A data frame is a great way to store your data, but accessing individual columns can become tedious. For a data frame called suburbs that contains a column called pop, here is the naïve way to calculate the z-scores of pop:

```
> z <- (suburbs$pop - mean(suburbs$pop)) / sd(suburbs$pop)
```

Call me a whiner, but all that typing gets tedious. The with function lets you expose the columns of a data frame as distinct variables. It takes two arguments, a data frame and an expression to be evaluated. Inside the expression, you can refer to the data frame columns by their names:

```
> z <- with(suburbs, (pop - mean(pop)) / sd(pop))
```

That is useful for one-liners. If you will be working repeatedly with columns in your data frame, attach the data frame to your search list and the columns will become available as variables:

```
> attach(suburbs)
```

After the attach, the second item in the search list is the suburbs data frame:

```
> search()
 [1] ".GlobalEnv"        "suburbs"           "package:stats"
 [4] "package:graphics"  "package:grDevices" "package:utils"
 [7] "package:datasets"  "package:methods"   "Autoloads"
[10] "package:base"
```

Now we can refer to the columns of the data frame as if they were variables:

```
> z <- (pop - mean(pop)) / sd(pop)
```

When you are done, use a detach (with no arguments) to remove the second location in the search list:

```
> detach()
> search()
 [1] ".GlobalEnv"        "package:stats"     "package:graphics"
 [4] "package:grDevices" "package:utils"     "package:datasets"
 [7] "package:methods"   "Autoloads"         "package:base"
```

Observe that suburbs is no longer in the search list.

Attaching a data frame has a big quirk: R attaches a *temporary copy* of the data frame, which means that changes to the original data frame are hidden. In this session fragment, notice how changing the data frame does not change our view of the attached data:

```
> attach(suburbs)
> pop
 [1] 2853114   90352  171782   94487  102746  106221  147779   76031   70834
[10]   72616   74239   83048   67232   75386   63348   91452
> suburbs$pop <- 0         # Overwrite data frame contents
> pop                      # Hey! It seems nothing changed
 [1] 2853114   90352  171782   94487  102746  106221  147779   76031   70834
[10]   72616   74239   83048   67232   75386   63348   91452
> suburbs$pop             # Contents of data frame did indeed change
 [1] 0 0 0 0 0 0 0 0 0 0 0 0 0 0 0 0
```

Another source of confusion is that assigning values to the exposed names does not work as you might expect. In the following fragment, you might think we are scaling pop by 1,000 but we are actually creating a new local variable:

```
> attach(suburbs)
> pop
 [1] 2853114   90352  171782   94487  102746  106221  147779   76031   70834
[10]   72616   74239   83048   67232   75386   63348   91452
> pop <- pop / 1000       # Achtung! This is creating a local variable called "pop"
> ls()                    # We can see the new variable in our workspace
[1] "pop"       "suburbs"
```

```
> suburbs$pop                    # Original data is unchanged
 [1] 2853114   90352  171782   94487  102746  106221  147779   76031   70834
[10]   72616   74239   83048   67232   75386   63348   91452
```

5.32 Converting One Atomic Value into Another

Problem

You have a data value which has an atomic data type: character, complex, double, integer, or logical. You want to convert this value into one of the other atomic data types.

Solution

For each atomic data type, there is a function for converting values to that type. The conversion functions for atomic types include:

- as.character(x)
- as.complex(x)
- as.numeric(x) or as.double(x)
- as.integer(x)
- as.logical(x)

Discussion

Converting one atomic type into another is usually pretty simple. If the conversion works, you get what you would expect. If it does not work, you get NA:

```
> as.numeric(" 3.14 ")
[1] 3.14
> as.integer(3.14)
[1] 3
> as.numeric("foo")
[1] NA
Warning message:
NAs introduced by coercion
> as.character(101)
[1] "101"
```

If you have a vector of atomic types, these functions apply themselves to every value. So the preceding examples of converting scalars generalize easily to converting entire vectors:

```
> as.numeric(c("1","2.718","7.389","20.086"))
[1]  1.000  2.718  7.389 20.086
> as.numeric(c("1","2.718","7.389","20.086", "etc."))
[1]  1.000  2.718  7.389 20.086     NA
Warning message:
NAs introduced by coercion
```

```
> as.character(101:105)
[1] "101" "102" "103" "104" "105"
```

When converting logical values into numeric values, R converts FALSE to 0 and
TRUE to 1:

```
> as.numeric(FALSE)
[1] 0
> as.numeric(TRUE)
[1] 1
```

This behavior is useful when you are counting occurrences of TRUE in vectors of logical
values. If logvec is a vector of logical values, then sum(logvec) does an implicit conver-
sion from logical to integer and returns the number of TRUEs.

5.33 Converting One Structured Data Type into Another

Problem

You want to convert a variable from one structured data type to another—for example,
converting a vector into a list or a matrix into a data frame.

Solution

These functions convert their argument into the corresponding structured data type:

- as.data.frame(x)
- as.list(x)
- as.matrix(x)
- as.vector(x)

Some of these conversions may surprise you, however. I suggest you review Table 5-1.

Discussion

Converting between structured data types can be tricky. Some conversions behave as
you'd expect. If you convert a matrix into a data frame, for instance, the rows and
columns of the matrix become the rows and columns of the data frame. No sweat.

Table 5-1. Data conversions

Conversion	How	Notes
Vector→List	as.list(vec)	Don't use list(vec); that creates a 1-element list whose only element is a copy of vec.
Vector→Matrix	To create a 1-column matrix: cbind(vec) *or* as.matrix(vec) To create a 1-row matrix: rbind(vec)	See Recipe 5.14.

Conversion	How	Notes
	To create an $n \times m$ matrix: `matrix(vec,n,m)`	
Vector→Data frame	To create a 1-column data frame: `as.data.frame(vec)`	
	To create a 1-row data frame: `as.data.frame(rbind(vec))`	
List→Vector	`unlist(lst)`	Use `unlist` rather than `as.vector`; see Note 1 and Recipe 5.11.
List→Matrix	To create a 1-column matrix: `as.matrix(lst)`	
	To create a 1-row matrix: `as.matrix(rbind(lst))`	
	To create an $n \times m$ matrix: `matrix(lst,n,m)`	
List→Data frame	If the list elements are columns of data: `as.data.frame(lst)`	
	If the list elements are rows of data: Recipe 5.19	
Matrix→Vector	`as.vector(mat)`	Returns all matrix elements in a vector.
Matrix→List	`as.list(mat)`	Returns all matrix elements in a list.
Matrix→Data frame	`as.data.frame(mat)`	
Data frame→Vector	To convert a 1-row data frame: `dfrm[1,]`	See Note 2.
	To convert a 1-column data frame: `dfrm[,1]` or `dfrm[[1]]`	
Data frame→List	`as.list(dfrm)`	See Note 3.
Data frame→Matrix	`as.matrix(dfrm)`	See Note 4.

In other cases, the results might surprise you. Table 5-1 summarizes some noteworthy examples. The following Notes are cited in that table:

1. When you convert a list into a vector, the conversion works cleanly if your list contains atomic values that are all of the same mode. Things become complicated if either (a) your list contains mixed modes (e.g., numeric and character), in which case everything is converted to characters; or (b) your list contains other structured data types, such as sublists or data frames—in which case very odd things happen, so don't do that.

2. Converting a data frame into a vector makes sense only if the data frame contains one row or one column. To extract all its elements into one, long vector, use `as.vector(as.matrix(dfrm))`. But even that makes sense only if the data frame is all-numeric or all-character; if not, everything is first converted to character strings.

3. Converting a data frame into a list may seem odd in that a data frame is already a list (i.e., a list of columns). Using `as.list` essentially removes the class (`data.frame`) and thereby exposes the underlying list. That is useful when you want R to treat your data structure as a list—say, for printing.

4. Be careful when converting a data frame into a matrix. If the data frame contains only numeric values then you get a numeric matrix. If it contains only character values, you get a character matrix. But if the data frame is a mix of numbers, characters, and/or factors, then all values are first converted to characters. The result is a matrix of character strings.

Problems with matrices

The matrix conversions detailed here assume that your matrix is homogeneous: all elements have the same mode (e.g, all numeric or all character). A matrix can to be heterogeneous, too, when the matrix is built from a list. If so, conversions become messy. For example, when you convert a mixed-mode matrix to a data frame, the data frame's columns are actually lists (to accommodate the mixed data).

See Also

See Recipe 5.32 for converting atomic data types; see the "Introduction" to this chapter for remarks on problematic conversions.

Data Transformations

Introduction

This chapter is all about the *apply* functions: `apply`, `lapply`, `sapply`, `tapply`, `mapply`; and their cousins, `by` and `split`. These functions let you take data in great gulps and process the whole gulp at once. Where traditional programming languages use loops, R uses vectorized operations and the apply functions to crunch data in batches, greatly streamlining the calculations.

Defining Groups Via a Factor

An important idiom of R is using a factor to define a group. Suppose we have a vector and a factor, both of the same length, that were created as follows:

```
> v <- c(40,2,83,28,58)
> f <- factor(c("A","C","C","B","C"))
```

We can visualize the vector elements and factors levels side by side, like this:

Vector	Factor
40	A
2	C
83	A
28	B
58	C

The factor level identifies the group of each vector element: 40 and 83 are in group A; 28 is in group B; and 2 and 58 are in group C.

In this book, I refer to such factors as *grouping factors*. They effectively slice and dice our data by putting them into groups. This is powerful because processing data in groups occurs often in statistics when comparing group means, comparing group proportions, performing ANOVA analysis, and so forth.

This chapter has recipes that use grouping factors to split vector elements into their respective groups (Recipe 6.1), apply a function to groups within a vector (Recipe 6.5), and apply a function to groups of rows within a data frame (Recipe 6.6). In other chapters, the same idiom is used to test group means (Recipe 9.19), perform one-way ANOVA analysis (Recipe 11.20), and plot data points by groups (Recipe 10.4), among other uses.

6.1 Splitting a Vector into Groups

Problem

You have a vector. Each element belongs to a different group, and the groups are identified by a grouping factor. You want to split the elements into the groups.

Solution

Suppose the vector is x and the factor is f. You can use the split function:

```
> groups <- split(x, f)
```

Alternatively, you can use the unstack function:

```
> groups <- unstack(data.frame(x,f))
```

Both functions return a list of vectors, where each vector contains the elements for one group.

The unstack function goes one step further: if all vectors have the same length, it converts the list into a data frame.

Discussion

The Cars93 dataset contains a factor called Origin that has two levels, USA and non-USA. It also contains a column called MPG.city. We can split the MPG data according to origin as follows:

```
> library(MASS)
> split(Cars93$MPG.city, Cars93$Origin)
$USA
 [1] 22 19 16 19 16 16 25 25 19 21 18 15 17 17 20 23 20 29 23 22 17 21 18 29 20
[26] 31 23 22 22 24 15 21 18 17 18 23 19 24 23 18 19 23 31 23 19 19 19 28

$non-USA
 [1] 25 18 20 19 22 46 30 24 42 24 29 22 26 20 17 18 18 29 28 26 18 17 20 19 29
[26] 18 29 24 17 21 20 33 25 23 39 32 25 22 18 25 17 21 18 21 20
```

The usefulness of splitting is that we can analyze the data by group. This example computes the median MPG for each group:

```
> g <- split(Cars93$MPG.city, Cars93$Origin)
> median(g[[1]])
[1] 20
> median(g[[2]])
[1] 22
```

See Also

See the "Introduction" to this chapter for more about grouping factors. Recipe 6.5 shows another way to apply functions, such as `median`, to each group. The `unstack` function can perform other, more powerful transformations beside this; see the help page.

6.2 Applying a Function to Each List Element

Problem

You have a list, and you want to apply a function to each element of the list.

Solution

Use either the `lapply` function or the `sapply` function, depending upon the desired form of the result. `lapply` always returns the results in list, whereas `sapply` returns the results in a vector if that is possible:

```
> lst <- lapply(lst, fun)
> vec <- sapply(lst, fun)
```

Discussion

These functions will call your function (*fun*, in the solution example) once for every element on your list. Your function should expect one argument, an element from the list. The `lapply` and `sapply` functions will collect the returned values. `lapply` collects them into a list and returns the list.

The "s" in "`sapply`" stands for "simplify." The function tries to simplify the results into a vector or matrix. For that to happen, all the returned values must have the same length. If that length is 1 then you get a vector; otherwise, you get a matrix. If the lengths vary, simplification is impossible and you get a list.

Let's say I teach an introductory statistics class four times and administer comparable final exams each time. Here are the exam scores from the four semesters:

```
> scores
$S1
 [1] 89 85 85 86 88 89 86 82 96 85 93 91 98 87 94 77 87 98 85 89
[21] 95 85 93 93 97 71 97 93 75 68 98 95 79 94 98 95

$S2
 [1]  60 98 94 95 99 97 100 73 93 91 98 86 66 83 77
```

```
[16]  97  91  93  71  91  95 100  72  96  91  76 100  97  99  95
[31]  97  77  94  99  88 100  94  93  86
```

$S3
```
 [1]  95  86  90  90  75  83  96  85  83  84  81  98  77  94  84  89  93  99  91  77
[21]  95  90  91  87  85  76  99  99  97  97  97  77  93  96  90  87  97  88
```

$S4
```
 [1]  67  93  63  83  87  97  96  92  93  96  87  90  94  90  82  91  85  93  83  90
[21]  87  99  94  88  90  72  81  93  93  94  97  89  96  95  82  97
```

Each semester starts with 40 students but, alas, not everyone makes it to the finish line; hence each semester has a different number of scores. We can count this number with the length function: lapply will return a list of lengths, and sapply will return a vector of lengths:

```
> lapply(scores, length)
$S1
[1] 36

$S2
[1] 39

$S3
[1] 38

$S4
[1] 36

> sapply(scores, length)
S1 S2 S3 S4
36 39 38 36
```

We can see the mean and standard deviation of the scores just as easily:

```
> sapply(scores, mean)
       S1       S2       S3       S4
88.77778 89.79487 89.23684 88.86111
> sapply(scores, sd)
       S1       S2       S3       S4
 7.720515 10.543592  7.178926  8.208542
```

If the called function returns a vector, sapply will form the results into a matrix. The range function, for example, returns a two-element vector:

```
> sapply(scores, range)
     S1  S2 S3 S4
[1,] 68  60 75 63
[2,] 98 100 99 99
```

If the called function returns a structured object, such as a list, then you will need to use lapply rather than sapply. Structured objects cannot be put into a vector. Suppose we want to perform a *t* test on every semester. The t.test function returns a list, so we must use lapply:

```
> tests <- lapply(scores, t.test)
```

Now the result tests is a list: it is a list of t.test results. We can use sapply to extract elements from the t.test results, such as the bounds of the confidence interval:

```
> sapply(tests, function(t) t$conf.int)
           S1       S2       S3       S4
[1,] 86.16553 86.37703 86.87719 86.08374
[2,] 91.39002 93.21271 91.59650 91.63848
```

See Also

See Recipe 2.12.

6.3 Applying a Function to Every Row

Problem

You have a matrix. You want to apply a function to every row, calculating the function result for each row.

Solution

Use the apply function. Set the second argument to 1 to indicate row-by-row application of a function:

```
> results <- apply(mat, 1, fun)      # mat is a matrix, fun is a function
```

The apply function will call fun once for each row, assemble the returned values into a vector, and then return that vector.

Discussion

Suppose your matrix long is longitudinal data. Each row contains data for one subject, and the columns contain the repeated observations over time:

```
> long
          trial1     trial2     trial3     trial4     trial5
Moe    -1.8501520 -1.406571 -1.0104817 -3.7170704 -0.2804896
Larry   0.9496313  1.346517 -0.1580926  1.6272786  2.4483321
Curly  -0.5407272 -1.708678 -0.3480616 -0.2757667 -1.2177024
```

You could calculate the average observation for each subject by applying the mean function to the rows. The result is a vector:

```
> apply(long, 1, mean)
        Moe      Larry      Curly
 -1.6529530  1.2427334 -0.8181872
```

Note that apply uses the rownames from your matrix to identify the elements of the resulting vector, which is handy.

The function being called (fun, described previously) should expect one argument, a vector, which will be one row from the matrix. The function can return a scalar or a vector. In the vector case, apply assembles the results into a matrix. The range function returns a vector of two elements, the minimum and the maximum, so applying it to long produces a matrix:

```
> apply(long, 1, range)
           Moe       Larry      Curly
[1,] -3.7170704 -0.1580926 -1.7086779
[2,] -0.2804896  2.4483321 -0.2757667
```

You can employ this recipe on data frames as well. It works if the data frame is homogeneous—either all numbers or all character strings. When the data frame has columns of different types, extracting vectors from the rows isn't sensible because vectors must be homogeneous.

6.4 Applying a Function to Every Column

Problem

You have a matrix or data frame, and you want to apply a function to every column.

Solution

For a matrix, use the apply function. Set the second argument to 2, which indicates column-by-column application of the function:

```
> results <- apply(mat, 2, fun)
```

For a data frame, use the lapply or sapply functions. Either one will apply a function to successive columns of your data frame. Their difference is that lapply assembles the return values into a list whereas sapply assembles them into a vector:

```
> lst <- lapply(dfrm, fun)
> vec <- sapply(dfrm, fun)
```

You can use apply on data frames, too, but only if the data frame is homogeneous (i.e., either all numeric values or all character strings).

Discussion

The apply function is intended for processing a matrix. In Recipe 6.3 we used apply to process the rows of a matrix. This is the same situation, but now we are processing the columns. The second argument of apply determines the direction:

- 1 means process row by row.
- 2 means process column by column.

This is more mnemonic than it looks. We speak of matrices in "rows and columns", so rows are first and columns second; 1 and 2, respectively.

A data frame is a more complicated data structure than a matrix, so there are more options. You can simply use apply, in which case R will convert your data frame to a matrix and then apply your function. That will work if your data frame contains only numeric data or character data. It probably will not work if you have mixed data types. In that case, R will force all columns to have identical types, likely performing an unwanted conversion as a result.

Fortunately, there is an alternative. Recall that a data frame is a kind of list: it is a list of the columns of the data frame. You can use lapply and sapply to process the columns, as described in Recipe 6.2:

```
> lst <- lapply(dfrm, fun)        # Returns a list
> vec <- sapply(dfrm, fun)        # Returns a vector
```

The function fun should expect one argument: a column from the data frame.

I often use this recipe to check the types of columns in data frames. The batch column of this data frame seems to contain numbers:

```
> head(batches)
  batch clinic dosage   shrinkage
1     1     IL      3 -0.11810714
2     3     IL      4 -0.29932107
3     2     IL      4 -0.27651716
4     1     IL      5 -0.18925825
5     2     IL      2 -0.06804804
6     3     NJ      5 -0.38279193
```

But printing the classes of the columns reveals it to be a factor instead:

```
> sapply(batches, class)
    batch    clinic    dosage shrinkage
 "factor"  "factor" "integer" "numeric"
```

A cool example of this recipe is removing low-correlation variables from a set of predictors. Suppose that resp is a response variable and pred is a data frame of predictor variables. Suppose further that we have too many predictors and therefore want to select the top 10 as measured by correlation with the response.

The first step is to calculate the correlation between each variable and resp. In R, that's a one-liner:

```
> cors <- sapply(pred, cor, y=resp)
```

The sapply function will call the function cor for every column in pred. Note that we gave a third argument, y=resp, to sapply. Any arguments beyond the second one are passed to cor. Every time that sapply calls cor, the first argument will be a column and the second argument will be y=resp. With those arguments, the function call will be cor(*column*,y=resp), which calculates the correlation between the given column and resp.

The result from `sapply` is a vector of correlations, one for each column. We use the `rank` function to find the positions of the correlations that have the largest magnitude:

```
> mask <- (rank(-abs(cors)) <= 10)
```

This expression is a comparison (`<=`), so it returns a vector of logical values. It is cleverly constructed so that the top 10 correlations have corresponding `TRUE` values and all others are `FALSE`. Using that vector of logical values, we can select just those columns from the data frame:

```
> best.pred <- pred[,mask]
```

At this point, we can regress `resp` against `best.pred`, knowing that we have chosen the predictors with the highest correlations:

```
> lm(resp ~ best.pred)
```

That's pretty good for four lines of code.

See Also

See Recipes 5.22, 6.2, and 6.3.

6.5 Applying a Function to Groups of Data

Problem

Your data elements occur in groups. You want to process the data by groups—for example, summing by group or averaging by group.

Solution

Create a grouping factor (of the same length as your vector) that identifies the group of each corresponding datum. Then use the `tapply` function, which will apply a function to each group of data:

```
> tapply(x, f, fun)
```

Here, *x* is a vector, *f* is a grouping factor, and *fun* is a function. The function should expect one argument, which is a vector of elements taken from *x* according to their group.

Discussion

Suppose I have a vector with the populations of the 16 largest cities in the greater Chicago metropolitan area, taken from the data frame called `suburbs`:

```
> attach(suburbs)
> pop
 [1] 2853114   90352  171782   94487  102746  106221  147779   76031   70834
[10]   72616   74239   83048   67232   75386   63348   91452
```

We can easily compute sums and averages for all the cities:

```
> sum(pop)
[1] 4240667
> mean(pop)
[1] 265041.7
```

What if we want the sum and average broken out by county? We will need a factor, say county, the same length as pop, where each level of the factor gives the corresponding county (there are two Lake counties: one in Illinois and one in Indiana) :

```
> county
 [1] Cook     Kenosha  Kane     Kane     Lake(IN) Kendall  DuPage   Cook
 [9] Will     Cook     Cook     Lake(IN) Cook     Cook     Cook     Lake(IL)
Levels: Cook DuPage Kane Kendall Kenosha Lake(IL) Lake(IN) Will
```

Now I can use the county factor and tapply function to process items in groups. The tapply function has three main parameters: the vector of data, the factor that defines the groups, and a function. It will extract each group, apply the function to each group, and return a vector with the collected results. This example shows summing the populations by county:

```
> tapply(pop,county,sum)
    Cook   DuPage     Kane  Kendall  Kenosha Lake(IL) Lake(IN)     Will
 3281966   147779   266269   106221    90352    91452   185794    70834
```

The next example computes average populations by county:

```
> tapply(pop,county,mean)
    Cook   DuPage     Kane  Kendall  Kenosha Lake(IL) Lake(IN)     Will
468852.3 147779.0 133134.5 106221.0  90352.0  91452.0  92897.0  70834.0
```

The function given to tapply should expect a single argument: a vector containing all the members of one group. A good example is the length function, which takes a vector parameter and returns the vector's length. Use it to count the number of data in each group; in this case, the number of cities in each county:

```
> tapply(pop,county,length)
    Cook   DuPage     Kane  Kendall  Kenosha Lake(IL) Lake(IN)     Will
       7        1        2        1        1        1        2        1
```

In most cases you will use functions that return a scalar and tapply will collect the returned scalars into a vector. Your function can return complex objects, too, in which case tapply will return them in a list. See the tapply help page for details.

See Also

See this chapter's "Introduction" for more about grouping factors.

6.6 Applying a Function to Groups of Rows

Problem

You want to apply a function to groups of rows within a data frame.

Solution

Define a grouping factor—that is, a factor with one level (element) for every row in your data frame—that identifies the data groups.

For each such group of rows, the by function puts the rows into a temporary data frame and calls your function with that argument. The by function collects the returned values into a list and returns the list:

```
> by(dfrm, fact, fun)
```

Here, *dfrm* is the data frame, *fact* is the grouping factor, and *fun* is a function. The function should expect one argument, a data frame.

Discussion

The advantage of the by function is that it calls your function with a data frame, which is useful if your function handles data frames in a special way. For instance, the print, summary, and mean functions perform special processing for data frames.

Suppose you have a data frame from clinical trials, called trials, where the dosage was randomized to study its effect:

```
> trials
    sex     pre dose1 dose2     post
1    F 5.931640     2     1 3.162600
2    F 4.496187     1     2 3.293989
3    M 6.161944     1     1 4.446643
4    F 4.322465     2     1 3.334748
5    M 4.153510     1     1 4.429382
.
. (etc.)
.
```

The data includes a factor for the subject's sex, so by can split the data according to sex and call summary for the two groups. The result is two summaries, one for men and one for women:

```
> by(trials, trials$sex, summary)
trials$sex: F
 sex         pre            dose1           dose2            post
 F:7   Min.   :4.156   Min.   :1.000   Min.   :1.000   Min.   :2.886
 M:0   1st Qu.:4.409   1st Qu.:1.000   1st Qu.:1.000   1st Qu.:3.075
       Median :4.895   Median :1.000   Median :2.000   Median :3.163
       Mean   :5.020   Mean   :1.429   Mean   :1.571   Mean   :3.174
       3rd Qu.:5.668   3rd Qu.:2.000   3rd Qu.:2.000   3rd Qu.:3.314
       Max.   :5.932   Max.   :2.000   Max.   :2.000   Max.   :3.389
```

```
----------------------------------------------------------------
trials$sex: M
 sex        pre              dose1            dose2            post
 F:0   Min.   :3.998    Min.   :1.000    Min.   :1.000    Min.   :3.738
 M:9   1st Qu.:4.773    1st Qu.:1.000    1st Qu.:1.000    1st Qu.:3.800
       Median :5.110    Median :2.000    Median :1.000    Median :4.194
       Mean   :5.189    Mean   :1.556    Mean   :1.444    Mean   :4.148
       3rd Qu.:5.828    3rd Qu.:2.000    3rd Qu.:2.000    3rd Qu.:4.429
       Max.   :6.658    Max.   :2.000    Max.   :2.000    Max.   :4.517
```

We can also build linear models of post as a function of the dosages, with one model for men and one model for women:

```
> models <- by(trials, trials$sex, function(df) lm(post~pre+dose1+dose2, data=df))
```

Observe that the parameter to our function is a data frame, so we can use it as the data argument of lm. The result is a two-element list of linear models. When we print the list, we see a model for each sex:

```
> print(models)
trials$sex: F

Call:
lm(formula = post ~ pre + dose1 + dose2, data = df)

Coefficients:
(Intercept)          pre         dose1         dose2
    4.30804     -0.08161      -0.16225      -0.31354

----------------------------------------------------------
trials$sex: M

Call:
lm(formula = post ~ pre + dose1 + dose2, data = df)

Coefficients:
(Intercept)          pre         dose1         dose2
    5.29981     -0.02713      -0.36851      -0.30323
```

We have a list of models, so we can apply the confint function to each list element and see the confidence intervals for each model's coefficients:

```
> lapply(models, confint)
$F
                  2.5 %        97.5 %
(Intercept)   3.0841733    5.53191431
pre          -0.2950747    0.13184560
dose1        -0.4711773    0.14667409
dose2        -0.6044273   -0.02264593

$M
                  2.5 %        97.5 %
(Intercept)   4.8898433    5.70978218
pre          -0.1070276    0.05277108
dose1        -0.4905828   -0.24644057
dose2        -0.4460200   -0.16043211
```

In this case, we see that pre is not significant for either model because its confidence interval contains zero; in contrast, dose1 and dose2 are significant for both models. The key fact, however, is that the models have significantly different intercepts, alerting us to a potentially different response for men and women.

See Also

See this chapter's "Introduction" for more about grouping factors. See Recipe 6.2 for more about lapply.

6.7 Applying a Function to Parallel Vectors or Lists

Problem

You have a function, say f, that takes multiple arguments. You want to apply the function element-wise to vectors and obtain a vector result. Unfortunately, the function is not *vectorized*; that is, it works on scalars but not on vectors.

Solution

Use the mapply function. It will apply the function f to your arguments element-wise:

```
> mapply(f, vec₁, vec₂, ..., vec_N)
```

There must be one vector for each argument expected by f. If the vector arguments are of unequal length, the Recycling Rule is applied.

The mapply function also works with list arguments:

```
> mapply(f, list₁, list₂, ..., list_N)
```

Discussion

The basic operators of R, such as $x + y$, are vectorized; this means that they compute their result element-by-element and return a vector of results. Also, many R functions are vectorized.

Not all functions are vectorized, however, and those that are not work only on scalars. Using vector arguments produces errors at best and meaningless results at worst. In such cases, the mapply function can effectively vectorize the function for you.

Consider the gcd function from Recipe 2.12, which takes two arguments:

```
> gcd <- function(a,b) {
+     if (b == 0) return(a)
+     else return(gcd(b, a %% b))
+ }
```

If we apply gcd to two vectors, the result is wrong answers and a pile of error messages:

```
> gcd(c(1,2,3), c(9,6,3))
[1] 1 2 0
Warning messages:
1: In if (b == 0) return(a) else return(gcd(b, a%%b)) :
  the condition has length > 1 and only the first element will be used
2: In if (b == 0) return(a) else return(gcd(b, a%%b)) :
  the condition has length > 1 and only the first element will be used
3: In if (b == 0) return(a) else return(gcd(b, a%%b)) :
  the condition has length > 1 and only the first element will be used
```

The function is not vectorized, but we can use mapply to vectorize it. This gives the element-wise GCDs between two vectors:

```
> mapply(gcd, c(1,2,3), c(9,6,3))
[1] 1 2 3
```

Strings and Dates

Introduction

Strings? Dates? In a statistical programming package?

As soon as you read files or print reports, you need strings. When you work with real-world problems, you need dates.

R has facilities for both strings and dates. They are clumsy compared to string-oriented languages such as Perl, but then it's a matter of the right tool for the job. I wouldn't want to perform logistic regression in Perl.

Classes for Dates and Times

R has a variety of classes for working with dates and times; which is nice if you prefer having a choice but annoying if you prefer living simply. There is a critical distinction among the classes: some are date-only classes, some are datetime classes. All classes can handle calendar dates (e.g., March 15, 2010), but not all can represent a datetime (11:45 AM on March 1, 2010).

The following classes are included in the base distribution of R:

Date

> The Date class can represent a calendar date but not a clock time. It is a solid, general-purpose class for working with dates, including conversions, formatting, basic date arithmetic, and time-zone handling. Most of the date-related recipes in this book are built on the Date class.

POSIXct

> This is a datetime class, and it can represent a moment in time with an accuracy of one second. Internally, the datetime is stored as the number of seconds since January 1, 1970, and so is a very compact representation. This class is recommended for storing datetime information (e.g., in data frames).

POSIXlt

This is also a datetime class, but the representation is stored in a nine-element list that includes the year, month, day, hour, minute, and second. That representation makes it easy to extract date parts, such as the month or hour. Obviously, this representation is much less compact than the POSIXct class; hence it is normally used for intermediate processing and not for storing data.

The base distribution also provides functions for easily converting between representations: as.Date, as.POSIXct, and as.POSIXlt.

The following packages are available for downloading from CRAN:

chron

The chron package can represent both dates and times but without the added complexities of handling time zones and daylight savings time. It's therefore easier to use than Date but less powerful than POSIXct and POSIXlt. It would be useful for work in econometrics or time series analysis.

lubridate

This is a relatively new, general-purpose package. It's designed to make working with dates and times easier while keeping the important bells and whistles such as time zones. It's especially clever regarding datetime arithmetic.

mondate

This is a specialized package for handling dates in units of months in addition to days and years. Such needs arise in accounting and actuarial work, for example, where month-by-month calculations are needed.

timeDate

This is a high-powered package with well-thought-out facilities for handling dates and times, including date arithmetic, business days, holidays, conversions, and generalized handling of time zones. It was originally part of the Rmetrics software for financial modeling, where precision in dates and times is critical. If you have a demanding need for date facilities, consider this package.

Which class should you select? The article "Date and Time Classes in R" by Grothendieck and Petzoldt offers this general advice:

> When considering which class to use, always choose the least complex class that will support the application. That is, use Date if possible, otherwise use chron and otherwise use the POSIX classes. Such a strategy will greatly reduce the potential for error and increase the reliability of your application.

See Also

See help(DateTimeClasses) for more details regarding the built-in facilities. See the June 2004 article "Date and Time Classes in R" (*http://cran.r-project.org/doc/Rnews/Rnews_2004-1.pdf*) by Gabor Grothendieck and Thomas Petzoldt for a great introduction to the date and time facilities. The June 2001 article "Date-Time Classes" (*http://cran.r*

-project.org/doc/Rnews/Rnews_2001-2.pdf) by Brian Ripley and Kurt Hornik discusses the two POSIX classes in particular.

7.1 Getting the Length of a String

Problem

You want to know the length of a string.

Solution

Use the nchar function, not the length function.

Discussion

The nchar function takes a string and returns the number of characters in the string:

```
> nchar("Moe")
[1] 3
> nchar("Curly")
[1] 5
```

If you apply nchar to a vector of strings, it returns the length of each string:

```
> s <- c("Moe", "Larry", "Curly")
> nchar(s)
[1] 3 5 5
```

You might think the length function returns the length of a string. Nope—it returns the length of a *vector*. When you apply the length function to a single string, R returns the value 1 because it views that string as a singleton vector—a vector with one element:

```
> length("Moe")
[1] 1
> length(c("Moe","Larry","Curly"))
[1] 3
```

7.2 Concatenating Strings

Problem

You want to join together two or more strings into one string.

Solution

Use the paste function.

Discussion

The `paste` function concatenates several strings together. In other words, it creates a new string by joining the given strings end to end:

```
> paste("Everybody", "loves", "stats.")
[1] "Everybody loves stats."
```

By default, `paste` inserts a single space between pairs of strings, which is handy if that's what you want and annoying otherwise. The `sep` argument lets you specify a different separator. Use an empty string (`""`) to run the strings together without separation:

```
> paste("Everybody", "loves", "stats.", sep="-")
[1] "Everybody-loves-stats."
> paste("Everybody", "loves", "stats.", sep="")
[1] "Everybodylovesstats."
```

The function is very forgiving about nonstring arguments. It tries to convert them to strings using the `as.character` function:

```
> paste("The square root of twice pi is approximately", sqrt(2*pi))
[1] "The square root of twice pi is approximately 2.506628274631"
```

If one or more arguments are vectors of strings, `paste` will generate all combinations of the arguments:

```
> stooges <- c("Moe", "Larry", "Curly")
> paste(stooges, "loves", "stats.")
[1] "Moe loves stats."   "Larry loves stats." "Curly loves stats."
```

Sometimes you want to join even those combinations into one, big string. The `collapse` parameter lets you define a top-level separator and instructs `paste` to concatenate the generated strings using that separator:

```
> paste(stooges, "loves", "stats", collapse=", and ")
[1] "Moe loves stats, and Larry loves stats, and Curly loves stats"
```

7.3 Extracting Substrings

Problem

You want to extract a portion of a string according to position.

Solution

Use substr(*string*,*start*,*end*) to extract the substring that begins at *start* and ends at *end*.

Discussion

The `substr` function takes a string, a starting point, and an ending point. It returns the substring between the starting to ending points:

```
> substr("Statistics", 1, 4)    # Extract first 4 characters
[1] "Stat"
> substr("Statistics", 7, 10)   # Extract last 4 characters
[1] "tics"
```

Just like many R functions, substr lets the first argument be a vector of strings. In that case, it applies itself to every string and returns a vector of substrings:

```
> ss <- c("Moe", "Larry", "Curly")
> substr(ss, 1, 3)              # Extract first 3 characters of each string
[1] "Moe" "Lar" "Cur"
```

In fact, all the arguments can be vectors, in which case substr will treat them as parallel vectors. From each string, it extracts the substring delimited by the corresponding entries in the starting and ending points. This can facilitate some useful tricks. For example, the following code snippet extracts the last two characters from each string; each substring starts on the penultimate character of the original string and ends on the final character:

```
> cities <- c("New York, NY", "Los Angeles, CA", "Peoria, IL")
> substr(cities, nchar(cities)-1, nchar(cities))
[1] "NY" "CA" "IL"
```

You can extend this trick into mind-numbing territory by exploiting the Recycling Rule, but I suggest you avoid the temptation.

7.4 Splitting a String According to a Delimiter

Problem

You want to split a string into substrings. The substrings are separated by a delimiter.

Solution

Use strsplit, which takes two arguments—the string and the delimiter of the substrings:

```
> strsplit(string, delimiter)
```

The *delimiter* can be either a simple string or a regular expression.

Discussion

It is common for a string to contain multiple substrings separated by the same delimiter. One example is a file path, whose components are separated by slashes (/):

```
> path <- "/home/mike/data/trials.csv"
```

We can split that path into its components by using strsplit with a delimiter of /:

```
> strsplit(path, "/")
[[1]]
[1] ""          "home"      "mike"      "data"      "trials.csv"
```

Notice that the first "component" is actually an empty string because nothing preceded the first slash.

Also notice that `strsplit` returns a list and that each element of the list is a vector of substrings. This two-level structure is necessary because the first argument can be a vector of strings. Each string is split into its substrings (a vector); then those vectors are returned in a list. This example splits three file paths and returns a three-element list:

```
> paths <- c("/home/mike/data/trials.csv",
+            "/home/mike/data/errors.csv",
+            "/home/mike/corr/reject.doc")
> strsplit(paths, "/")
[[1]]
[1] ""          "home"      "mike"      "data"      "trials.csv"

[[2]]
[1] ""          "home"      "mike"      "data"      "errors.csv"

[[3]]
[1] ""          "home"      "mike"      "corr"      "reject.doc"
```

The second argument of `strsplit` (the *delimiter* argument) is actually much more powerful than these examples indicate. It can be a regular expression, letting you match patterns far more complicated than a simple string. In fact, to defeat the regular expression feature (and its interpretation of special characters) you must include the `fixed=TRUE` argument.

See Also

To learn more about regular expressions in R, see the help page for `regexp`. See O'Reilly's *Mastering Regular Expressions* (*http://oreilly.com/catalog/9780596528126/*), by Jeffrey E.F. Friedl to learn more about regular expressions in general.

7.5 Replacing Substrings

Problem

Within a string, you want to replace one substring with another.

Solution

Use `sub` to replace the first instance of a substring:

```
> sub(old, new, string)
```

Use `gsub` to replace all instances of a substring:

```
> gsub(old, new, string)
```

Discussion

The sub function finds the first instance of the *old* substring within *string* and replaces it with the *new* substring:

```
> s <- "Curly is the smart one. Curly is funny, too."
> sub("Curly", "Moe", s)
[1] "Moe is the smart one. Curly is funny, too."
```

gsub does the same thing, but it replaces all instances of the substring (a global replace), not just the first:

```
> gsub("Curly", "Moe", s)
[1] "Moe is the smart one. Moe is funny, too."
```

To remove a substring altogether, simply set the *new* substring to be empty:

```
> sub(" and SAS", "", "For really tough problems, you need R and SAS.")
[1] "For really tough problems, you need R."
```

The *old* argument can be regular expression, which allows you to match patterns much more complicated than a simple string. This is actually assumed by default, so you must set the fixed=TRUE argument if you don't want sub and gsub to interpret *old* as a regular expression.

See Also

To learn more about regular expressions in R, see the help page for regexp. See *Mastering Regular Expressions* to learn more about regular expressions in general.

7.6 Seeing the Special Characters in a String

Problem

Your string contains special characters that are unprintable. You want to know what they are.

Solution

Use print to see the special characters in a string. The cat function will not reveal them.

Discussion

I once encountered a bizarre situation in which a string contained, unbeknownst to me, unprintable characters. My R script used cat to display the string, and the script output was garbled until I realized there were unprintable characters in the string.

In this example, the string seems to contain 13 characters but `cat` shows only a 6-character output:

```
> nchar(s)
[1] 13
> cat(s)
second
```

The reason is that the string contains *special characters*, which are characters that do not display when printed. When we use `print`, it uses escapes (backslashes) to show the special characters:

```
> print(s)
[1] "first\rsecond\n"
```

Notice that the string contains a return character (\r) in the middle, so "first" is overwritten by "second" when `cat` displays the string.

7.7 Generating All Pairwise Combinations of Strings

Problem

You have two sets of strings, and you want to generate all combinations from those two sets (their Cartesian product).

Solution

Use the `outer` and `paste` functions together to generate the matrix of all possible combinations:

```
> m <- outer(strings1, strings2, paste, sep="")
```

Discussion

The `outer` function is intended to form the outer product. However, it allows a third argument to replace simple multiplication with any function. In this recipe we replace multiplication with string concatenation (`paste`), and the result is all combinations of strings.

Suppose you have four test sites and three treatments:

```
> locations <- c("NY", "LA", "CHI", "HOU")
> treatments <- c("T1", "T2", "T3")
```

We can apply `outer` and `paste` to generate all combinations of test sites and treatments:

```
> outer(locations, treatments, paste, sep="-")
     [,1]      [,2]      [,3]
[1,] "NY-T1"   "NY-T2"   "NY-T3"
[2,] "LA-T1"   "LA-T2"   "LA-T3"
[3,] "CHI-T1"  "CHI-T2"  "CHI-T3"
[4,] "HOU-T1"  "HOU-T2"  "HOU-T3"
```

The fourth argument of outer is passed to paste. In this case, we passed sep="-" in order to define a hyphen as the separator between the strings.

The result of outer is a matrix. If you want the combinations in a vector instead, flatten the matrix using the as.vector function.

In the special case when you are combining a set with itself and order does not matter, the result will be duplicate combinations:

```
> outer(treatments, treatments, paste, sep="-")
     [,1]    [,2]    [,3]
[1,] "T1-T1" "T1-T2" "T1-T3"
[2,] "T2-T1" "T2-T2" "T2-T3"
[3,] "T3-T1" "T3-T2" "T3-T3"
```

But suppose we want all *unique* pairwise combinations of treatments. We can eliminate the duplicates by removing the lower triangle (or upper triangle). The lower.tri function identifies that triangle, so inverting it identifies all elements *outside* the lower triangle:

```
> m <- outer(treatments, treatments, paste, sep="-")
> m[!lower.tri(m)]
[1] "T1-T1" "T1-T2" "T2-T2" "T1-T3" "T2-T3" "T3-T3"
```

See Also

See Recipe 7.2 for using paste to generate combinations of strings.

7.8 Getting the Current Date

Problem

You need to know today's date.

Solution

The Sys.Date function returns the current date:

```
> Sys.Date()
[1] "2010-02-11"
```

Discussion

The Sys.Date function returns a Date object. In the preceding example it seems to return a string because the result is printed inside double quotes. What really happened, however, is that Sys.Date returned a Date object and then R converted that object into a string for printing purposes. You can see this by checking the class of the result from Sys.Date:

```
> class(Sys.Date())
[1] "Date"
```

See Also

See Recipe 7.10.

7.9 Converting a String into a Date

Problem

You have the string representation of a date, such as "2010-12-31", and you want to convert that into a Date object.

Solution

You can use as.Date, but you must know the format of the string. By default, as.Date assumes the string looks like *yyyy-mm-dd*. To handle other formats, you must specify the format parameter of as.Date. Use format="%m/%d/%Y" if the date is in American style, for instance.

Discussion

This example shows the default format assumed by as.Date, which is the ISO 8601 standard format of *yyyy-mm-dd*:

```
> as.Date("2010-12-31")
[1] "2010-12-31"
```

The as.Date function returns a Date object that is here being converted back to a string for printing; this explains the double quotes around the output.

The string can be in other formats, but you must provide a format argument so that as.Date can interpret your string. See the help page for the stftime function for details about allowed formats.

Being a simple American, I often mistakenly try to convert the usual American date format (*mm/dd/yyyy*) into a Date object, with these unhappy results:

```
> as.Date("12/31/2010")
Error in charToDate(x) :
  character string is not in a standard unambiguous format
```

Here is the correct way to convert an American-style date:

```
> as.Date("12/31/2010", format="%m/%d/%Y")
[1] "2010-12-31"
```

Observe that the Y in the format string is capitalized to indicate a 4-digit year. If you're using 2-digit years, specify a lowercase y.

7.10 Converting a Date into a String

Problem

You want to convert a Date object into a character string, usually because you want to print the date.

Solution

Use either format or as.character:

```
> format(Sys.Date())
[1] "2010-04-01"
> as.character(Sys.Date())
[1] "2010-04-01"
```

Both functions allow a format argument that controls the formatting. Use format="%m/%d/%Y" to get American-style dates, for example:

```
> format(Sys.Date(), format="%m/%d/%Y")
[1] "04/01/2010"
```

Discussion

The format argument defines the appearance of the resulting string. Normal characters, such as slash (/) or hyphen (-) are simply copied to the output string. Each two-letter combination of a percent sign (%) followed by another character has special meaning. Some common ones are:

%b

 Abbreviated month name ("Jan")

%B

 Full month name ("January")

%d

 Day as a two-digit number

%m

 Month as a two-digit number

%y

 Year without century (00–99)

%Y

 Year with century

See the help page for the strftime function for a complete list of formatting codes.

7.11 Converting Year, Month, and Day into a Date

Problem

You have a date represented by its year, month, and day. You want to merge these elements into a single Date object representation.

Solution

Use the ISOdate function:

```
> ISOdate(year, month, day)
```

The result is a POSIXct object that you can convert into a Date object:

```
> as.Date(ISOdate(year, month, day))
```

Discussion

It is common for input data to contain dates encoded as three numbers: year, month, and day. The ISOdate function can combine them into a POSIXct object:

```
> ISOdate(2012,2,29)
[1] "2012-02-29 12:00:00 GMT"
```

You can keep your date in the POSIXct format. However, when working with pure dates (not dates and times), I often convert to a Date object and truncate the unused time information:

```
> as.Date(ISOdate(2012,2,29))
[1] "2012-02-29"
```

Trying to convert an invalid date results in NA:

```
> ISOdate(2013,2,29)        # Oops! 2013 is not a leap year
[1] NA
```

ISOdate can process entire vectors of years, months, and days, which is quite handy for mass conversion of input data. The following example starts with the year/month/day numbers for the third Wednesday in January of several years and then combines them all into Date objects:

```
> years
[1] 2010 2011 2012 2013 2014
> months
[1] 1 1 1 1 1
> days
[1] 15 21 20 18 17
> ISOdate(years, months, days)
[1] "2010-01-15 12:00:00 GMT" "2011-01-21 12:00:00 GMT"
[3] "2012-01-20 12:00:00 GMT" "2013-01-18 12:00:00 GMT"
[5] "2014-01-17 12:00:00 GMT"
> as.Date(ISOdate(years, months, days))
[1] "2010-01-15" "2011-01-21" "2012-01-20" "2013-01-18" "2014-01-17"
```

Purists will note that the vector of months is redundant and that the last expression can therefore be further simplified by invoking the Recycling Rule:

```
> as.Date(ISOdate(years, 1, days))
[1] "2010-01-15" "2011-01-21" "2012-01-20" "2013-01-18" "2014-01-17"
```

This recipe can also be extended to handle year, month, day, hour, minute, and second data by using the ISOdatetime function (see the help page for details):

```
> ISOdatetime(year, month, day, hour, minute, second)
```

7.12 Getting the Julian Date

Problem

Given a Date object, you want to extract the Julian date—which is, in R, the number of days since January 1, 1970.

Solution

Either convert the Date object to an integer or use the julian function:

```
> d <- as.Date("2010-03-15")
> as.integer(d)
[1] 14683
> julian(d)
[1] 14683
attr(,"origin")
[1] "1970-01-01"
```

Discussion

A Julian "date" is simply the number of days since a more-or-less arbitrary starting point. In the case of R, that starting point is January 1, 1970, the same starting point as Unix systems. So the Julian date for January 1, 1970 is zero, as shown here:

```
> as.integer(as.Date("1970-01-01"))
[1] 0
> as.integer(as.Date("1970-01-02"))
[1] 1
> as.integer(as.Date("1970-01-03"))
[1] 2
.
. (etc.)
.
```

7.13 Extracting the Parts of a Date

Problem

Given a Date object, you want to extract a date part such as the day of the week, the day of the year, the calendar day, the calendar month, or the calendar year.

Solution

Convert the Date object to a POSIXlt object, which is a list of date parts. Then extract the desired part from that list:

```
> d <- as.Date("2010-03-15")
> p <- as.POSIXlt(d)
> p$mday                    # Day of the month
[1] 15
> p$mon                     # Month (0 = January)
[1] 2
> p$year + 1900             # Year
[1] 2010
```

Discussion

The POSIXlt object represents a date as a list of date parts. Convert your Date object to POSIXlt by using the as.POSIXlt function, which will give you a list with these members:

sec
 Seconds (0–61)

min
 Minutes (0–59)

hour
 Hours (0–23)

mday
 Day of the month (1–31)

mon
 Month (0–11)

year
 Years since 1900

wday
 Day of the week (0–6, 0 = Sunday)

yday
 Day of the year (0–365)

isdst
 Daylight savings time flag

Using these date parts, we can learn that April 1, 2010, is a Thursday (wday = 4) and the 91st day of the year (because yday = 0 on January 1):

```
> d <- as.Date("2010-04-01")
> as.POSIXlt(d)$wday
[1] 4
> as.POSIXlt(d)$yday
[1] 90
```

A common mistake is failing to add 1900 to the year, giving the impression you are living a long, long time ago:

```
> as.POSIXlt(d)$year            # Oops!
[1] 110
> as.POSIXlt(d)$year + 1900
[1] 2010
```

7.14 Creating a Sequence of Dates

Problem

You want to create a sequence of dates, such as a sequence of daily, monthly, or annual dates.

Solution

The seq function is a generic function that has a version for Date objects. It can create a Date sequence similarly to the way it creates a sequence of numbers.

Discussion

A typical use of seq specifies a starting date (from), ending date (to), and increment (by). An increment of 1 indicates daily dates:

```
> s <- as.Date("2012-01-01")
> e <- as.Date("2012-02-01")
> seq(from=s, to=e, by=1)                # One month of dates
 [1] "2012-01-01" "2012-01-02" "2012-01-03" "2012-01-04" "2012-01-05" "2012-01-06"
 [7] "2012-01-07" "2012-01-08" "2012-01-09" "2012-01-10" "2012-01-11" "2012-01-12"
[13] "2012-01-13" "2012-01-14" "2012-01-15" "2012-01-16" "2012-01-17" "2012-01-18"
[19] "2012-01-19" "2012-01-20" "2012-01-21" "2012-01-22" "2012-01-23" "2012-01-24"
[25] "2012-01-25" "2012-01-26" "2012-01-27" "2012-01-28" "2012-01-29" "2012-01-30"
[31] "2012-01-31" "2012-02-01"
```

Another typical use specifies a starting date (from), increment (by), and number of dates (length.out):

```
> seq(from=s, by=1, length.out=7)        # Dates, one week apart
[1] "2012-01-01" "2012-01-02" "2012-01-03" "2012-01-04" "2012-01-05" "2012-01-06"
[7] "2012-01-07"
```

The increment (by) is flexible and can be specified in days, weeks, months, or years:

```
> seq(from=s, by="month", length.out=12)        # First of the month for one year
[1] "2012-01-01" "2012-02-01" "2012-03-01" "2012-04-01" "2012-05-01" "2012-06-01"
[7] "2012-07-01" "2012-08-01" "2012-09-01" "2012-10-01" "2012-11-01" "2012-12-01"
> seq(from=s, by="3 months", length.out=4)       # Quarterly dates for one year
[1] "2012-01-01" "2012-04-01" "2012-07-01" "2012-10-01"
> seq(from=s, by="year", length.out=10)          # Year-start dates for one decade
[1] "2012-01-01" "2013-01-01" "2014-01-01" "2015-01-01" "2016-01-01" "2017-01-01"
[7] "2018-01-01" "2019-01-01" "2020-01-01" "2021-01-01"
```

Be careful with by="month" near month-end. In this example, the end of February overflows into March, which is probably not what you wanted:

```
> seq(as.Date("2010-01-29"), by="month", len=3)
[1] "2010-01-29" "2010-03-01" "2010-03-29"
```

Probability

Introduction

Probability theory is the foundation of statistics, and R has plenty of machinery for working with probability, probability distributions, and random variables. The recipes in this chapter show you how to calculate probabilities from quantiles, calculate quantiles from probabilities, generate random variables drawn from distributions, plot distributions, and so forth.

Names of Distributions

R has an abbreviated name for every probability distribution. This name is used to identify the functions associated with the distribution. For example, the name of the Normal distribution is "norm", which is the root of these function names:

Function	Purpose
dnorm	Normal density
pnorm	Normal distribution function
qnorm	Normal quantile function
rnorm	Normal random variates

Table 8-1 describes some common discrete distributions, and Table 8-2 describes several common continuous distributions.

Table 8-1. Discrete distributions

Discrete distribution	R name	Parameters
Binomial	binom	n = number of trials; p = probability of success for one trial
Geometric	geom	p = probability of success for one trial
Hypergeometric	hyper	m = number of white balls in urn; n = number of black balls in urn; k = number of balls drawn from urn

Discrete distribution	R name	Parameters
Negative binomial (NegBinomial)	nbinom	size = number of successful trials; either prob = probability of successful trial or mu = mean
Poisson	pois	lambda = mean

Table 8-2. Continuous distributions

Continuous distribution	R name	Parameters
Beta	beta	shape1; shape2
Cauchy	cauchy	location; scale
Chi-squared (Chisquare)	chisq	df = degrees of freedom
Exponential	exp	rate
F	f	df1 and df2 = degrees of freedom
Gamma	gamma	rate; either rate or scale
Log-normal (Lognormal)	lnorm	meanlog = mean on logarithmic scale; sdlog = standard deviation on logarithmic scale
Logistic	logis	location; scale
Normal	norm	mean; sd = standard deviation
Student's *t* (TDist)	t	df = degrees of freedom
Uniform	unif	min = lower limit; max = upper limit
Weibull	weibull	shape; scale
Wilcoxon	wilcox	m = number of observations in first sample; n = number of observations in second sample

All distribution-related functions require distributional parameters, such as `size` and `prob` for the binomial or `prob` for the geometric. The big "gotcha" is that the distributional parameters may not be what you expect. For example, I would expect the parameter of an exponential distribution to be β, the mean. The R convention, however, is for the exponential distribution to be defined by the *rate* $= 1/\beta$, so I often supply the wrong value. The moral is, study the help page before you use a function related to a distribution. Be sure you've got the parameters right.

Getting Help on Probability Distributions

To see the R functions related to a particular probability distribution, use the help command and the full name of the distribution. For example, this will show the functions related to the Normal distribution:

```
> ?Normal
```

Some distributions have names that don't work well with the help command, such as "Student's t". They have special help names, as noted in Tables 8-1 and 8-2: NegBinomial, Chisquare, Lognormal, and TDist. Thus, to get help on the Student's t distribution, use this:

```
> ?TDist
```

See Also

There are many other distributions implemented in downloadable packages; see the CRAN task view devoted to probability distributions (*http://cran.r-project.org/web/views/Distributions.html*). The SuppDists package is part of the R base, and it includes ten supplemental distributions. The MASS package, which is also part of the base, provides additional support for distributions, such as maximum-likelihood fitting for some common distributions as well as sampling from a multivariate Normal distribution.

8.1 Counting the Number of Combinations

Problem

You want to calculate the number of combinations of n items taken k at a time.

Solution

Use the choose function:

```
> choose(n, k)
```

Discussion

A common problem in computing probabilities of discrete variables is *counting combinations*: the number of distinct subsets of size k that can be created from n items. The number is given by $n!/r!(n - r)!$, but it's much more convenient to use the choose function—especially as n and k grow larger:

```
> choose(5,3)       # How many ways can we select 3 items from 5 items?
[1] 10
> choose(50,3)      # How many ways can we select 3 items from 50 items?
[1] 19600
> choose(50,30)     # How many ways can we select 30 items from 50 items?
[1] 4.712921e+13
```

These numbers are also known as *binomial coefficients*.

See Also

This recipe merely counts the combinations; see Recipe 8.2 to actually generate them.

8.2 Generating Combinations

Problem

You want to generate all combinations of *n* items taken *k* at a time.

Solution

Use the combn function:

```
> combn(items, k)
```

Discussion

We can use combn(1:5,3) to generate all combinations of the numbers 1 through 5 taken three at a time:

```
> combn(1:5,3)
     [,1] [,2] [,3] [,4] [,5] [,6] [,7] [,8] [,9] [,10]
[1,]    1    1    1    1    1    1    2    2    2     3
[2,]    2    2    2    3    3    4    3    3    4     4
[3,]    3    4    5    4    5    5    4    5    5     5
```

The function is not restricted to numbers. We can generate combinations of strings, too. Here are all combinations of five treatments taken three at a time:

```
> combn(c("T1","T2","T3","T4","T5"), 3)
     [,1] [,2] [,3] [,4] [,5] [,6] [,7] [,8] [,9] [,10]
[1,] "T1" "T1" "T1" "T1" "T1" "T1" "T2" "T2" "T2" "T3"
[2,] "T2" "T2" "T2" "T3" "T3" "T4" "T3" "T3" "T4" "T4"
[3,] "T3" "T4" "T5" "T4" "T5" "T5" "T4" "T5" "T5" "T5"
```

 As the number of items, *n*, increases, the number of combinations can explode—especially if *k* is not near to 1 or *n*.

See Also

See Recipe 8.1 to count the number of possible combinations *before* you generate a huge set.

8.3 Generating Random Numbers

Problem

You want to generate random numbers.

Solution

The simple case of generating a uniform random number between 0 and 1 is handled by the `runif` function. This example generates one uniform random number:

```
> runif(1)
```

R can generate random variates from other distributions, however. For a given distribution, the name of the random number generator is "r" prefixed to the distribution's abbreviated name (e.g., `rnorm` for the Normal distribution's random number generator). This example generates one random value from the standard normal distribution:

```
> rnorm(1)
```

Discussion

Most programming languages have a wimpy random number generator that generates one random number, uniformly distributed between 0.0 and 1.0, and that's all. Not R.

R can generate random numbers from many probability distributions other than the uniform distribution. The simple case of generating uniform random numbers between 0 and 1 is handled by the `runif` function:

```
> runif(1)
[1] 0.5119812
```

The argument of `runif` is the number of random values to be generated. Generating a vector of 10 such values is as easy as generating one:

```
> runif(10)
 [1] 0.03475948 0.88950680 0.90005434 0.95689496 0.55829493 0.18407604
 [7] 0.87814788 0.71057726 0.11140864 0.66392239
```

There are random number generators for all built-in distributions. Simply prefix the distribution name with "r" and you have the name of the corresponding random number generator. Here are some common ones:

```
> runif(1, min=-3, max=3)     # One uniform variate between -3 and +3
[1] 2.954591
> rnorm(1)                     # One standard Normal variate
[1] 1.048491
> rnorm(1, mean=100, sd=15)    # One Normal variate, mean 100 and SD 15
[1] 108.7300
> rbinom(1, size=10, prob=0.5) # One binomial variate
[1] 3
> rpois(1, lambda=10)          # One Poisson variate
[1] 13
> rexp(1, rate=0.1)            # One exponential variate
[1] 8.430267
> rgamma(1, shape=2, rate=0.1) # One gamma variate
[1] 20.47334
```

As with `runif`, the first argument is the number of random values to be generated. Subsequent arguments are the parameters of the distribution, such as `mean` and `sd` for the Normal distribution or `size` and `prob` for the binomial. See the function's R help page for details.

The examples given so far use simple scalars for distributional parameters. Yet the parameters can also be vectors, in which case R will cycle through the vector while generating random values. The following example generates three normal random values drawn from distributions with means of –10, 0, and +10, respectively (all distributions have a standard deviation of 1.0):

```
> rnorm(3, mean=c(-10,0,+10), sd=1)
[1] -11.195667   2.615493  10.294831
```

That is a powerful capability in such cases as hierarchical models, where the parameters are themselves random. The next example calculates 100 draws of a normal variate whose mean is itself randomly distributed and with hyperparameters of $\mu = 0$ and $\sigma = 0.2$:

```
> means <- rnorm(100, mean=0, sd=0.2)
> rnorm(100, mean=means, sd=1)
  [1] -0.410299581  1.030662055 -0.920933054 -0.572994026  0.984743043
  [6]  1.251189879  0.873930251  0.509027943 -0.788626886 -0.113224062
 [11] -0.035042586  0.150925067  0.634036678  1.627473761 -0.812021925
.
. (etc.)
.
```

If you are generating many random values and the vector of parameters is too short, R will apply the Recycling Rule to the parameter vector.

See Also

See the "Introduction" to this chapter.

8.4 Generating Reproducible Random Numbers

Problem

You want to generate a sequence of random numbers, but you want to reproduce the same sequence every time your program runs.

Solution

Before running your R code, call the `set.seed` function to initialize the random number generator to a known state:

```
> set.seed(666)        # Or use any other positive integer
```

Discussion

After generating random numbers, you may often want to reproduce the same sequence of "random" numbers every time your program executes. That way, you get the same results from run to run. I once supported a complicated Monte Carlo analysis of a huge portfolio of securities. The users complained about getting a slightly different result each time the program ran. No kidding, I replied; it's all based on random numbers, so of course there is randomness in the output. The solution was to set the random number generator to a known state at the beginning of the program. That way, it would generate the same (quasi-)random numbers each time and thus yield consistent, reproducible results.

In R, the set.seed function sets the random number generator to a known state. The function takes one argument, an integer. Any positive integer will work, but you must use the same one in order to get the same initial state.

The function returns nothing. It works behind the scenes, initializing (or reinitializing) the random number generator. The key here is that using the same seed restarts the random number generator back at the same place:

```
> set.seed(165)         # Initialize the random number generator to a known state
> runif(10)             # Generate ten random numbers
 [1] 0.1159132 0.4498443 0.9955451 0.6106368 0.6159386 0.4261986 0.6664884
 [8] 0.1680676 0.7878783 0.4421021
> set.seed(165)         # Reinitialize to the same known state
> runif(10)             # Generate the same ten "random" numbers
 [1] 0.1159132 0.4498443 0.9955451 0.6106368 0.6159386 0.4261986 0.6664884
 [8] 0.1680676 0.7878783 0.4421021
```

 When you set the seed value and freeze your sequence of random numbers, you are eliminating a source of randomness that is critical to algorithms such as Monte Carlo simulations. Before you call set.seed in your application, ask yourself: Am I undercutting the value of my program or perhaps even damaging its logic?

See Also

See Recipe 8.3 for more about generating random numbers.

8.5 Generating a Random Sample

Problem

You want to sample a dataset randomly.

Solution

The `sample` function will randomly select *n* items from a vector:

```
> sample(vec, n)
```

Discussion

Suppose your World Series data contains a vector of years when the Series was played. You can select 10 years at random using `sample`:

```
> sample(world.series$year, 10)
 [1] 1906 1963 1966 1928 1905 1924 1961 1959 1927 1934
```

The items are randomly selected, so running `sample` again (usually) produces a different result:

```
> sample(world.series$year, 10)
 [1] 1968 1947 1966 1916 1970 1961 1936 1913 1914 1958
```

The `sample` function normally samples *without replacement*, meaning it will not select the same item twice. Some statistical procedures (especially the bootstrap) require sampling *with* replacement, which means that one item can appear multiple times in the sample. Specify `replace=TRUE` to sample with replacement.

It's easy to implement a simple bootstrap using sampling with replacement. This code fragment repeatedly samples a dataset x and calculates the sample median:

```
medians <- numeric(1000)
for (i in 1:1000) {
      medians[i] <- median(sample(x, replace=TRUE))
}
```

From the bootstrap estimates, we can estimate the confidence interval for the median:

```
ci <- quantile(medians, c(0.025, 0.975))
cat("95% confidence interval is (", ci, ")\n")
```

Typical output would be:

```
95% confidence interval is ( 1.642021 1.702843 )
```

See Also

See Recipe 8.7 for randomly permuting a vector and Recipe 13.8 for more about bootstrapping.

8.6 Generating Random Sequences

Problem

You want to generate a random sequence, such as a series of simulated coin tosses or a simulated sequence of Bernoulli trials.

Solution

Use the `sample` function. Sample from the set of possible values, and set `replace=TRUE`:

```
> sample(set, n, replace=TRUE)
```

Discussion

The `sample` function randomly selects items from a set. It normally samples *without* replacement, which means that it will not select the same item twice. With `replace=TRUE`, however, `sample` can select items over and over; this allows you to generate long, random sequences of items.

The following example generates a random sequence of 10 simulated flips of a coin:

```
> sample(c("H","T"), 10, replace=TRUE)
 [1] "H" "H" "H" "T" "T" "H" "T" "H" "H" "T"
```

The next example generates a sequence of 20 *Bernoulli trials*—random successes or failures. We use TRUE to signify a success:

```
> sample(c(FALSE,TRUE), 20, replace=TRUE)
 [1]  TRUE FALSE FALSE FALSE FALSE FALSE FALSE  TRUE  TRUE  TRUE  TRUE  TRUE
[13]  TRUE  TRUE FALSE  TRUE  TRUE FALSE FALSE  TRUE
```

By default, `sample` will choose equally among the set elements and so the probability of selecting either TRUE or FALSE is 0.5. With a Bernoulli trial, the probability p of success is not necessarily 0.5. You can bias the sample by using the `prob` argument of `sample`; this argument is a vector of probabilities, one for each set element. Suppose we want to generate 20 Bernoulli trials with a probability of success $p = 0.8$. We set the probability of FALSE to be 0.2 and the probability of TRUE to 0.8:

```
> sample(c(FALSE,TRUE), 20, replace=TRUE, prob=c(0.2,0.8))
 [1]  TRUE  TRUE  TRUE  TRUE  TRUE  TRUE  TRUE FALSE  TRUE  TRUE  TRUE  TRUE
[13]  TRUE  TRUE  TRUE  TRUE  TRUE  TRUE FALSE  TRUE
```

The resulting sequence is clearly biased toward TRUE. I chose this example because it's a simple demonstration of a general technique. For the special case of a binary-valued sequence you can use `rbinom`, the random generator for binomial variates:

```
> rbinom(10, 1, 0.8)
 [1] 1 0 1 1 1 1 1 1 1 1
```

8.7 Randomly Permuting a Vector

Problem

You want to generate a random permutation of a vector.

Solution

If `v` is your vector, then `sample(v)` returns a random permutation.

Discussion

I associate the `sample` function with sampling from large datasets. However, the default parameters enable you to create a random rearrangement of the dataset. The function call `sample(v)` is equivalent to:

```
sample(v, size=length(v), replace=FALSE)
```

which means "select all the elements of `v` in random order while using each element exactly once. That is a random permutation." Here is a random permutation of 1, ..., 10:

```
> sample(1:10)
 [1]  5  8  7  4  3  9  2  6  1 10
```

See Also

See Recipe 8.5 for more about `sample`.

8.8 Calculating Probabilities for Discrete Distributions

Problem

You want to calculate either the simple or the cumulative probability associated with a discrete random variable.

Solution

For a simple probability, $P(X = x)$, use the density function. All built-in probability distributions have a density function whose name is "d" prefixed to the distribution name—for example, `dbinom` for the binomial distribution.

For a cumulative probability, $P(X \leq x)$, use the distribution function. All built-in probability distributions have a distribution function whose name is "p" prefixed to the distribution name; thus, `pbinom` is the distribution function for the binomial distribution.

Discussion

Suppose we have a binomial random variable X over 10 trials, where each trial has a success probability of 1/2. Then we can calculate the probability of observing $x = 7$ by calling `dbinom`:

```
> dbinom(7, size=10, prob=0.5)
[1] 0.1171875
```

That calculates a probability of about 0.117. R calls `dbinom` the *density function*. Some textbooks call it the *probability mass function* or the *probability function*. Calling it a density function keeps the terminology consistent between discrete and continuous distributions (Recipe 8.9).

The cumulative probability, $P(X \le x)$, is given by the *distribution function*, which is sometimes called the *cumulative probability function*. The distribution function for the binomial distribution is pbinom. Here is the cumulative probability for $x = 7$ (i.e., $P(X \le 7)$):

```
> pbinom(7, size=10, prob=0.5)
[1] 0.9453125
```

It appears the probability of observing $X \le 7$ is about 0.945.

Here are the density functions and distribution functions for some common discrete distributions:

Distribution	Density function: $P(X = x)$	Distribution function: $P(X \le x)$
Binomial	dbinom(x, size, prob)	pbinom(x, size, prob)
Geometric	dgeom(x, prob)	pgeom(x, prob)
Poisson	dpois(x, lambda)	ppois(x, lambda)

The complement of the cumulative probability is the *survival function*, $P(X > x)$. All of the distribution functions let you find this right-tail probability simply by specifying lower.tail=FALSE:

```
> pbinom(7, size=10, prob=0.5, lower.tail=FALSE)
[1] 0.0546875
```

Thus we see that the probability of observing $X > 7$ is about 0.055.

The interval probability, $P(x_1 < X \le x_2)$, is the probability of observing X between the limits x_1 and x_2. It is simply calculated as the difference between two cumulative probabilities: $P(X \le x_2) - P(X \le x_1)$. Here is $P(3 < X \le 7)$ for our binomial variable:

```
> pbinom(7,size=10,prob=0.5) - pbinom(3,size=10,prob=0.5)
[1] 0.7734375
```

R lets you specify multiple values of x for these functions and will return a vector of the corresponding probabilities. Here we calculate two cumulative probabilities, $P(X \le 3)$ and $P(X \le 7)$, in one call to pbinom:

```
> pbinom(c(3,7), size=10, prob=0.5)
[1] 0.1718750 0.9453125
```

This leads to a one-liner for calculating interval probabilities. The diff function calculates the difference between successive elements of a vector. We apply it to the output of pbinom to obtain the difference in cumulative probabilities—in other words, the interval probability:

```
> diff(pbinom(c(3,7), size=10, prob=0.5))
[1] 0.7734375
```

See Also

See this chapter's "Introduction" for more about the built-in probability distributions.

8.9 Calculating Probabilities for Continuous Distributions

Problem

You want to calculate the *distribution function* (DF) or *cumulative distribution function* (CDF) for a continuous random variable.

Solution

Use the distribution function, which calculates $P(X \leq x)$. All built-in probability distributions have a distribution function whose name is "p" prefixed to the distribution's abbreviated name—for instance, pnorm for the Normal distribution.

Discussion

The R functions for probability distributions follow a consistent pattern, so the solution to this recipe is essentially identical to the solution for discrete random variables (Recipe 8.8). The significant difference is that continuous variables have no "probability" at a single point, $P(X = x)$. Instead, they have a density at a point.

Given that consistency, the discussion of distribution functions in Recipe 8.8 is applicable here, too. The following table gives the distribution function for several continuous distributions:

Distribution	Distribution function: $P(X \leq x)$
Normal	pnorm(x, mean, sd)
Student's t	pt(x, df)
Exponential	pexp(x, rate)
Gamma	pgamma(x, shape, rate)
Chi-squared (χ^2)	pchisq(x, df)

We can use pnorm to calculate the probability that a man is shorter than 66 inches, assuming that men's heights are normally distributed with a mean of 70 inches and a standard deviation of 3 inches. Mathematically speaking, we want $P(X \leq 66)$ given that $X \sim N(70, 3)$:

```
> pnorm(66, mean=70, sd=3)
[1] 0.09121122
```

Likewise, we can use pexp to calculate the probability that an exponential variable with a mean of 40 could be less than 20:

```
> pexp(20, rate=1/40)
[1] 0.3934693
```

Just as for discrete probabilities, the functions for continuous probabilities use lower.tail=FALSE to specify the survival function, $P(X > x)$. This call to pexp gives the probability that the same exponential variable could be greater than 50:

```
> pexp(50, rate=1/40, lower.tail=FALSE)
[1] 0.2865048
```

Also like discrete probabilities, the interval probability for a continuous variable, $P(x_1 < X < x_2)$, is computed as the difference between two cumulative probabilities, $P(X < x_2) - P(X < x_1)$. For the same exponential variable, here is $P(20 < X < 50)$, the probability that it could fall between 20 and 50:

```
> pexp(50,rate=1/40) - pexp(20,rate=1/40)
[1] 0.3200259
```

See Also

See this chapter's "Introduction" for more about the built-in probability distributions.

8.10 Converting Probabilities to Quantiles

Problem

Given a probability p and a distribution, you want to determine the corresponding *quantile* for p: the value x such that $P(X \le x) = p$.

Solution

Every built-in distribution includes a quantile function that converts probabilities to quantiles. The function's name is "q" prefixed to the distribution name; thus, for instance, qnorm is the quantile function for the Normal distribution.

The first argument of the quantile function is the probability. The remaining arguments are the distribution's parameters, such as mean, shape, or rate:

```
> qnorm(0.05, mean=100, sd=15)
[1] 75.3272
```

Discussion

A common example of computing quantiles is when we compute the limits of a confidence interval. If we want to know the 95% confidence interval ($\alpha = 0.05$) of a standard normal variable, then we need the quantiles with probabilities of $\alpha/2 = 0.025$ and $(1 - \alpha)/2 = 0.975$:

```
> qnorm(0.025)
[1] -1.959964
> qnorm(0.975)
[1] 1.959964
```

In the true spirit of R, the first argument of the quantile functions can be a vector of probabilities, in which case we get a vector of quantiles. We can simplify this example into a one-liner:

```
> qnorm(c(0.025,0.975))
[1] -1.959964  1.959964
```

All the built-in probability distributions provide a quantile function. Here are the quantile functions for some common discrete distributions:

Distribution	Quantile function
Binomial	qbinom(p, size, prob)
Geometric	qgeom(p, prob)
Poisson	qpois(p, lambda)

And here are the quantile functions for common continuous distributions:

Distribution	Quantile function
Normal	qnorm(p, mean, sd)
Student's t	qt(p, df)
Exponential	qexp(p, rate)
Gamma	qgamma(p, shape, rate=$rate$) or qgamma(p, shape, scale=$scale$)
Chi-squared (χ^2)	qchisq(p, df)

See Also

Determining the quantiles of a data set is different from determining the quantiles of a distribution—see Recipe 9.5.

8.11 Plotting a Density Function

Problem

You want to plot the density function of a probability distribution.

Solution

Define a vector x over the domain. Apply the distribution's density function to x and then plot the result. This code snippet plots the standard normal distribution:

```
> x <- seq(from=-3, to=+3, length.out=100)
> plot(x, dnorm(x))
```

Discussion

All the built-in probability distributions include a density function. For a particular
density, the function name is "d" prepended to the density name. The density function
for the Normal distribution is dnorm, the density for the gamma distribution is dgamma,
and so forth.

If the first argument of the density function is a vector, then the function calculates the
density at each point and returns the vector of densities.

The following code creates a 2 × 2 plot of four densities, as shown in Figure 8-1:

```
x <- seq(from=0, to=6, length.out=100)          # Define the density domains
ylim <- c(0, 0.6)

par(mfrow=c(2,2))                                # Create a 2x2 plotting area

plot(x, dunif(x,min=2,max=4), main="Uniform",   # Plot a uniform density
     type='l', ylim=ylim)
plot(x, dnorm(x,mean=3,sd=1), main="Normal",     # Plot a Normal density
     type='l', ylim=ylim)
plot(x, dexp(x,rate=1/2), main="Exponential",    # Plot an exponential density
     type='l', ylim=ylim)
plot(x, dgamma(x,shape=2,rate=1), main="Gamma",  # Plot a gamma density
     type='l', ylim=ylim)
```

A raw density plot is rarely useful or interesting by itself, and we usually shade a region
of interest. Figure 8-2 shows a standard normal distribution with shading for the region
of $1 \le z \le 2$.

We create the plot by first plotting the density and then shading a region with the
polygon function. The polygon function draws a series of lines around the highlighted
area and fills it.

First, we draw the density curve:

```
x <- seq(from=-3, to=+3, length.out=100)
y <- dnorm(x)
plot(x, y, main="Standard Normal Distribution", type='l',
     ylab="Density", xlab="Quantile")
abline(h=0)
```

Next, we define the region of interest by a series of line segments. The line segments,
in turn, are defined by a series of (x, y) coordinates. The polygon function will connect
the first and last (x, y) points to close the polygon:

```
# The body of the polygon follows the density curve where 1 <= z <= 2
region.x <- x[1 <= x & x <= 2]
region.y <- y[1 <= x & x <= 2]

# We add initial and final segments, which drop down to the Y axis
```

```
region.x <- c(region.x[1], region.x, tail(region.x,1))
region.y <- c(        0, region.y,                0)
```

Finally, we call `polygon` to plot the boundary of the region and fill it:

```
polygon(region.x, region.y, density=10)
```

By default, `polygon` does not fill the region. Setting `density=10` turns on filling with thin lines at a 45° angle. To fill with a color instead, we would set `density` to –1 and set `col` to the desired color:

```
polygon(region.x, region.y, density=-1, col="red")
```

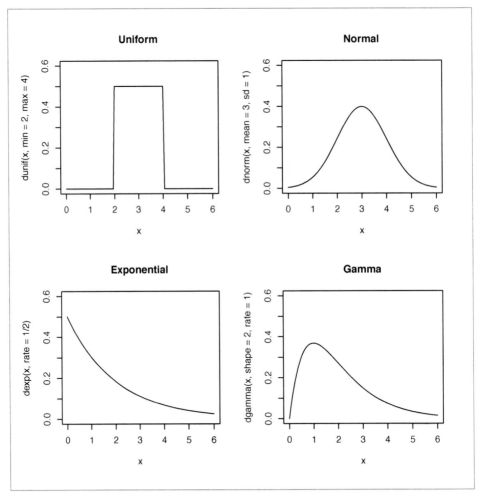

Figure 8-1. Plotting density functions

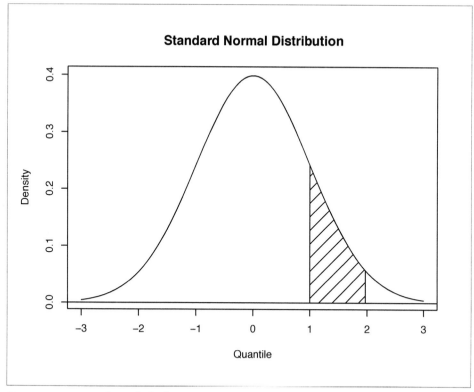

Figure 8-2. Density plot with shading

General Statistics

Introduction

Any significant application of R includes statistics or models or graphics. This chapter addresses the statistics. Some recipes simply describe how to calculate a statistic, such as relative frequency. Most recipes involve statistical tests or confidence intervals. The statistical tests let you choose between two competing hypotheses; that paradigm is described next. Confidence intervals reflect the likely range of a population parameter and are calculated based on your data sample.

Null Hypotheses, Alternative Hypotheses, and *p*-Values

Many of the statistical tests in this chapter use a time-tested paradigm of statistical inference. In the paradigm, we have one or two data samples. We also have two competing hypotheses, either of which could reasonably be true.

One hypothesis, called the *null hypothesis*, is that *nothing happened*: the mean was unchanged; the treatment had no effect; you got the expected answer; the model did not improve; and so forth.

The other hypothesis, called the *alternative hypothesis*, is that *something happened*: the mean rose; the treatment improved the patients' health; you got an unexpected answer; the model fit better; and so forth.

We want to determine which hypothesis is more likely in light of the data:

1. To begin, we assume that the null hypothesis is true.

2. We calculate a test statistic. It could be something simple, such as the mean of the sample, or it could be quite complex. The critical requirement is that we must know the statistic's distribution. We might know the distribution of the sample mean, for example, by invoking the Central Limit Theorem.

3. From the statistic and its distribution we can calculate a *p-value*, the probability of a test statistic value as extreme or more extreme than the one we observed, while assuming that the null hypothesis is true.

4. If the *p*-value is too small, we have strong evidence against the null hypothesis. This is called *rejecting the null hypothesis*.

5. If the *p*-value is not small then we have no such evidence. This is called *failing to reject the null hypothesis*.

There is one necessary decision here: When is a *p*-value "too small"?

 In this book, I follow the common convention that we reject the null hypothesis when $p < 0.05$ and fail to reject it when $p > 0.05$. In statistical terminology, I chose a *significance level* of $\alpha = 0.05$ to define the border between strong evidence and insufficient evidence against the null hypothesis.

But the real answer is, "it depends". Your chosen significance level depends on your problem domain. The conventional limit of $p < 0.05$ works for many problems. In my work, the data is especially noisy and so I am often satisfied with $p < 0.10$. For someone working in high-risk areas, $p < 0.01$ or $p < 0.001$ might be necessary.

In the recipes, I mention which tests include a *p*-value so that you can compare the *p*-value against your chosen significance level of α. I worded the recipes to help you interpret the comparison. Here is the wording from Recipe 9.4, a test for the independence of two factors:

> Conventionally, a *p*-value of less than 0.05 indicates that the variables are likely not independent whereas a *p*-value exceeding 0.05 fails to provide any such evidence.

This is a compact way of saying:

- The null hypothesis is that the variables are independent.
- The alternative hypothesis is that the variables are not independent.
- For $\alpha = 0.05$, if $p < 0.05$ then we reject the null hypothesis, giving strong evidence that the variables are not independent; if $p > 0.05$, we fail to reject the null hypothesis.
- You are free to choose your own α, of course, in which case your decision to reject or fail to reject might be different.

Remember, the recipe states the *informal interpretation* of the test results, not the rigorous mathematical interpretation. I use colloquial language in the hope that it will guide you toward a practical understanding and application of the test. If the precise semantics of hypothesis testing is critical for your work, I urge you to consult the reference cited under "See Also" on page 197 or one of the other fine textbooks on mathematical statistics.

Confidence Intervals

Hypothesis testing is a well-understood mathematical procedure, but it can be frustrating. First, the semantics is tricky. The test does not reach a definite, useful conclusion. You might get strong evidence against the null hypothesis, but that's all you'll get. Second, it does not give you a number, only evidence.

If you want numbers then use confidence intervals, which bound the estimate of a population parameter at a given level of confidence. Recipes in this chapter can calculate confidence intervals for means, medians, and proportions of a population.

For example, Recipe 9.9 calculates a 95% confidence interval for the population mean based on sample data. The interval is $97.16 < \mu < 103.98$, which means there is a 95% probability that the population's mean, μ, is between 97.16 and 103.98.

See Also

Statistical terminology and conventions can vary. This book generally follows the conventions of *Mathematical Statistics with Applications*, 6th ed., by Wackerly et al. (Duxbury Press). I recommend this book also for learning more about the statistical tests described in this chapter.

9.1 Summarizing Your Data

Problem

You want a basic statistical summary of your data.

Solution

The summary function gives some useful statistics for vectors, matrices, factors, and data frames:

```
> summary(vec)
   Min. 1st Qu.  Median    Mean 3rd Qu.    Max.
 0.1803  1.1090  1.9520  2.1750  2.6920  7.3290
```

Discussion

The Solution exhibits the summary of a vector. The 1st Qu. and 3rd Qu. are the first and third quartile, respectively. Having both the median and mean is useful because you can quickly detect skew. The Solution, for example, shows a mean that is larger than the median; this indicates a possible skew to the right.

The summary of a matrix works column by column. Here we see the summary of a matrix, mat, with three columns named Samp1, Samp2, and Samp3:

```
> summary(mat)
     Samp1             Samp2              Samp3
 Min.   :0.1803   Min.   : 0.1378   Min.   :0.1605
 1st Qu.:1.1094   1st Qu.: 1.2686   1st Qu.:1.0056
 Median :1.9517   Median : 1.8733   Median :1.8466
 Mean   :2.1748   Mean   : 2.2436   Mean   :2.0562
 3rd Qu.:2.6916   3rd Qu.: 2.9373   3rd Qu.:2.7666
 Max.   :7.3289   Max.   :11.4175   Max.   :6.3870
```

The summary of a factor gives counts:

```
> summary(fac)
  Yes    No Maybe
   44    20    36
```

The summary of a data frame incorporates all these features. It works column by column, giving an appropriate summary according to the column type. Numeric values receive a statistical summary and factors are counted (character strings are not summarized):

```
> summary(suburbs)
     city             county    state        pop
 Length:16        Cook    :7   IL:12   Min.   :  63348
 Class :character Kane    :2   IN: 2   1st Qu.:  73833
 Mode  :character Lake(IN):2   WI: 2   Median :  86700
                  DuPage  :1           Mean   : 265042
                  Kendall :1           3rd Qu.: 103615
                  Kenosha :1           Max.   :2853114
                  (Other) :2
```

The "summary" of a list is pretty funky: just the data type of each list member. Here is a summary of a list of vectors:

```
> summary(vec.list)
      Length Class  Mode
Samp1 100    -none- numeric
Samp2 100    -none- numeric
Samp3 100    -none- numeric
```

To summarize a list of vectors, apply summary to each list element:

```
> lapply(vec.list, summary)
$Samp1
   Min. 1st Qu.  Median    Mean 3rd Qu.    Max.
 0.1803  1.1090  1.9520  2.1750  2.6920  7.3290

$Samp2
   Min. 1st Qu.  Median    Mean 3rd Qu.    Max.
 0.1378  1.2690  1.8730  2.2440  2.9370 11.4200

$Samp3
   Min. 1st Qu.  Median    Mean 3rd Qu.    Max.
 0.1605  1.0060  1.8470  2.0560  2.7670  6.3870
```

Unfortunately, the summary function does not compute any measure of variability, such as standard deviation or median absolute deviation. This is a serious shortcoming, so I usually call sd or mad right after calling summary.

See Also

See Recipes 2.6 and 6.2.

9.2 Calculating Relative Frequencies

Problem

You want to count the relative frequency of certain observations in your sample.

Solution

Identify the interesting observations by using a logical expression; then use the mean function to calculate the fraction of observations it identifies. For example, given a vector x, you can find the relative frequency of positive values in this way:

```
> mean(x > 0)
```

Discussion

A logical expression, such as $x > 0$, produces a vector of logical values (TRUE and FALSE), one for each element of x. The mean function converts those values to 1s and 0s, respectively, and computes the average. This gives the fraction of values that are TRUE—in other words, the relative frequency of the interesting values. In the Solution, for example, that's the relative frequency of positive values.

The concept here is pretty simple. The tricky part is dreaming up a suitable logical expression. Here are some examples:

mean(lab == "NJ")
: Fraction of lab values that are New Jersey

mean(after > before)
: Fraction of observations for which the effect increases

mean(abs(x-mean(x)) > 2*sd(x))
: Fraction of observations that exceed two standard deviations from the mean

mean(diff(ts) > 0)
: Fraction of observations in a time series that are larger than the previous observation

9.3 Tabulating Factors and Creating Contingency Tables

Problem

You want to tabulate one factor or to build a contingency table from multiple factors.

Solution

The table function produces counts of one factor:

```
> table(f)
```

It can also produce contingency tables (cross-tabulations) from two or more factors:

```
> table(f1, f2)
```

Discussion

The table function counts the levels of one factor, such as these counts of initial and outcome (which are factors):

```
> table(initial)
initial
  Yes    No Maybe
   37    36    27
> table(outcome)
outcome
Fail Pass
  47   53
```

The greater power of table is in producing *contingency tables*, also known as *cross-tabulations*. Each cell in a contingency table counts how many times that row–column combination occurred:

```
> table(initial, outcome)
        outcome
initial Fail Pass
   Yes    13   24
   No     24   12
   Maybe  10   17
```

This table shows that the combination of initial = Yes and outcome = Fail occurred 13 times, the combination of initial = Yes and outcome = Pass occurred 24 times, and so forth.

See Also

The xtabs function can also produce a contingency table. It has a formula interface, which some people prefer.

9.4 Testing Categorical Variables for Independence

Problem

You have two categorical variables that are represented by factors. You want to test them for independence using the chi-squared test.

Solution

Use the `table` function to produce a contingency table from the two factors. Then use the `summary` function to perform a chi-squared test of the contingency table:

```
> summary(table(fac1,fac2))
```

The output includes a *p*-value. Conventionally, a *p*-value of less than 0.05 indicates that the variables are likely not independent whereas a *p*-value exceeding 0.05 fails to provide any such evidence.

Discussion

This example performs a chi-squared test on the contingency table of Recipe 9.3 and yields a *p*-value of 0.01255:

```
> summary(table(initial,outcome))
Number of cases in table: 100
Number of factors: 2
Test for independence of all factors:
        Chisq = 8.757, df = 2, p-value = 0.01255
```

The small *p*-value indicates that the two factors, `initial` and `outcome`, are probably not independent. Practically speaking, we conclude there is some connection between the variables.

See Also

The `chisq.test` function can also perform this test.

9.5 Calculating Quantiles (and Quartiles) of a Dataset

Problem

Given a fraction *f*, you want to know the corresponding quantile of your data. That is, you seek the observation *x* such that the fraction of observations below *x* is *f*.

Solution

Use the `quantile` function. The second argument is the fraction, *f*:

```
> quantile(vec, f)
```

For quartiles, simply omit the second argument altogether:

```
> quantile(vec)
```

Discussion

Suppose vec contains 1,000 observations between 0 and 1. The quantile function can tell you which observation delimits the lower 5% of the data:

```
> quantile(vec, .05)
      5%
0.04575003
```

The quantile documentation refers to the second argument as a "probability", which is natural when we think of probability as meaning relative frequency.

In true R style, the second argument can be a vector of probabilities; in this case, quantile returns a vector of corresponding quantiles, one for each probability:

```
> quantile(vec, c(.05, .95))
      5%        95%
0.04575003 0.95122306
```

That is a handy way to identify the middle 90% (in this case) of the observations.

If you omit the probabilities altogether then R assumes you want the probabilities 0, 0.25, 0.50, 0.75, and 1.0—in other words, the quartiles:

```
> quantile(vec)
         0%          25%          50%          75%         100%
0.001285589  0.260075658  0.479866042  0.734801500  0.997817661
```

Amazingly, the quantile function implements nine (yes, nine) different algorithms for computing quantiles. Study the help page before assuming that the default algorithm is the best one for you.

9.6 Inverting a Quantile

Problem

Given an observation x from your data, you want to know its corresponding quantile. That is, you want to know what fraction of the data is less than x.

Solution

Assuming your data is in a vector vec, compare the data against the observation and then use mean to compute the relative frequency of values less than x:

```
> mean(vec < x)
```

Discussion

The expression vec < x compares every element of vec against x and returns a vector of logical values, where the nth logical value is TRUE if vec[n] < x. The mean function converts those logical values to 0 and 1: 0 for FALSE and 1 for TRUE. The average of all those 1s and 0s is the fraction of vec that is less than x, or the inverse quantile of x.

See Also

This is an application of the general approach described in Recipe 9.2.

9.7 Converting Data to Z-Scores

Problem

You have a dataset, and you want to calculate the corresponding z-scores for all data elements. (This is sometimes called *normalizing the data*.)

Solution

Use the scale function:

```
> scale(x)
```

This works for vectors, matrices, and data frames. In the case of a vector, scale returns the vector of normalized values. In the case of matrices and data frames, scale normalizes each column independently and returns columns of normalized values in a matrix.

Discussion

You might also want to normalize a single value y relative to a dataset x. That can be done by using vectorized operations as follows:

```
> (y - mean(x)) / sd(x)
```

9.8 Testing the Mean of a Sample (*t* Test)

Problem

You have a sample from a population. Given this sample, you want to know if the mean of the population could reasonably be a particular value m.

Solution

Apply the t.test function to the sample x with the argument mu=m:

```
> t.test(x, mu=m)
```

The output includes a *p*-value. Conventionally, if $p < 0.05$ then the population mean is unlikely to be *m* whereas $p > 0.05$ provides no such evidence.

If your sample size *n* is small, then the underlying population must be normally distributed in order to derive meaningful results from the *t* test. A good rule of thumb is that "small" means $n < 30$.

Discussion

The *t* test is a workhorse of statistics, and this is one of its basic uses: making inferences about a population mean from a sample. The following example simulates sampling from a normal population with mean $\mu = 100$. It uses the *t* test to ask if the population mean could be 95, and t.test reports a *p*-value of 0.001897:

```
> x <- rnorm(50, mean=100, sd=15)
> t.test(x, mu=95)

        One Sample t-test

data:  x
t = 3.2832, df = 49, p-value = 0.001897
alternative hypothesis: true mean is not equal to 95
95 percent confidence interval:
  97.16167 103.98297
sample estimates:
mean of x
 100.5723
```

The *p*-value is small and so it's unlikely (based on the sample data) that 95 could be the mean of the population.

Informally, we could interpret the low *p*-value as follows. If the population mean were really 95, then the probability of observing our test statistic ($t = 3.2832$ or something more extreme) would be only 0.001897. That is very improbable, yet that is the value we observed. Hence we conclude that the null hypothesis is wrong; therefore, the sample data does not support the claim that the population mean is 95.

In sharp contrast, testing for a mean of 100 gives a *p*-value of 0.7374:

```
> t.test(x, mu=100)

        One Sample t-test

data:  x
t = 0.3372, df = 49, p-value = 0.7374
alternative hypothesis: true mean is not equal to 100
95 percent confidence interval:
  97.16167 103.98297
sample estimates:
mean of x
 100.5723
```

The large *p*-value indicates that the sample is consistent with assuming a population mean μ of 100. In statistical terms, the data does not provide evidence against the true mean being 100.

A common case is testing for a mean of zero. If you omit the mu argument, it defaults to zero.

See Also

The t.test is a many-splendored thing. See Recipes 9.9 and 9.15 for other uses.

9.9 Forming a Confidence Interval for a Mean

Problem

You have a sample from a population. Given that sample, you want to determine a confidence interval for the population's mean.

Solution

Apply the t.test function to your sample *x*:

```
> t.test(x)
```

The output includes a confidence interval at the 95% confidence level. To see intervals at other levels, use the conf.level argument.

As in Recipe 9.8, if your sample size *n* is small then the underlying population must be normally distributed for there to be a meaningful confidence interval. Again, a good rule of thumb is that "small" means $n < 30$.

Discussion

Applying the t.test function to a vector yields a lot of output. Buried in the output is a confidence interval:

```
> x <- rnorm(50, mean=100, sd=15)
> t.test(x)

        One Sample t-test

data:  x
t = 59.2578, df = 49, p-value < 2.2e-16
alternative hypothesis: true mean is not equal to 0
95 percent confidence interval:
  97.16167 103.98297
sample estimates:
mean of x
 100.5723
```

In this example, the confidence interval is approximately $97.16 < \mu < 103.98$, which is sometimes written simply as (97.16, 103.98).

We can raise the confidence level to 99% by setting `conf.level=0.99`:

```
> t.test(x, conf.level=0.99)

        One Sample t-test

data:  x
t = 59.2578, df = 49, p-value < 2.2e-16
alternative hypothesis: true mean is not equal to 0
99 percent confidence interval:
  96.0239 105.1207
sample estimates:
mean of x
 100.5723
```

That change widens the confidence interval to $96.02 < \mu < 105.12$.

9.10 Forming a Confidence Interval for a Median

Problem

You have a data sample, and you want to know the confidence interval for the median.

Solution

Use the `wilcox.test` function, setting `conf.int=TRUE`:

```
> wilcox.test(x, conf.int=TRUE)
```

The output will contain a confidence interval for the median.

Discussion

The procedure for calculating the confidence interval of a mean is well-defined and widely known. The same is not true for the median, unfortunately. There are several procedures for calculating the median's confidence interval. None of them is "the" procedure, as far as I know, but the Wilcoxon signed rank test is pretty standard.

The `wilcox.test` function implements that procedure. Buried in the output is the 95% confidence interval, which is approximately (0.424, 0.892) in this case:

```
> wilcox.test(x, conf.int=TRUE)

        Wilcoxon signed rank test

data:  x
V = 465, p-value = 1.863e-09
alternative hypothesis: true location is not equal to 0
95 percent confidence interval:
 0.4235421 0.8922106
```

```
sample estimates:
(pseudo)median
     0.6249297
```

You can change the confidence level by setting `conf.level`, such as `conf.level=0.99` or other such values.

The output also includes something called the *pseudomedian*, which is defined on the help page. Don't assume it equals the median; they are different:

```
> median(x)
[1] 0.547129
```

See Also

The bootstrap procedure is also useful for estimating the median's confidence interval; see Recipes 8.5 and 13.8.

9.11 Testing a Sample Proportion

Problem

You have a sample of values from a population consisting of successes and failures. You believe the true proportion of successes is p, and you want to test that hypothesis using the sample data.

Solution

Use the `prop.test` function. Suppose the sample size is n and the sample contains x successes:

```
> prop.test(x, n, p)
```

The output includes a *p*-value. Conventionally, a *p*-value of less than 0.05 indicates that the true proportion is unlikely to be p whereas a *p*-value exceeding 0.05 fails to provide such evidence.

Discussion

Suppose you encounter some loudmouthed fan of the Chicago Cubs early in the baseball season. The Cubs have played 20 games and won 11 of them, or 55% of their games. Based on that evidence, the fan is "very confident" that the Cubs will win more than half of their games this year. Should he be that confident?

The `prop.test` function can evaluate the fan's logic. Here, the number of observations is $n = 20$, the number of successes is $x = 11$, and p is the true probability of winning a game. We want to know whether it is reasonable to conclude, based on the data, that $p > 0.5$. Normally, `prop.test` would check for $p \neq 0.05$ but we can check for $p > 0.5$ instead by setting `alternative="greater"`:

```
> prop.test(11, 20, 0.5, alternative="greater")

        1-sample proportions test with continuity correction

data:  11 out of 20, null probability 0.5
X-squared = 0.05, df = 1, p-value = 0.4115
alternative hypothesis: true p is greater than 0.5
95 percent confidence interval:
 0.3496150 1.0000000
sample estimates:
   p
0.55
```

The prop.test output shows a large *p*-value, 0.4115, so we cannot reject the null hypothesis; that is, we cannot reasonably conclude that *p* is greater than 1/2. The Cubs fan is being overly confident based on too little data. No surprise there.

9.12 Forming a Confidence Interval for a Proportion

Problem

You have a sample of values from a population consisting of successes and failures. Based on the sample data, you want to form a confidence interval for the population's proportion of successes.

Solution

Use the prop.test function. Suppose the sample size is *n* and the sample contains *x* successes:

```
> prop.test(n, x)
```

The function output includes the confidence interval for *p*.

Discussion

I subscribe to a stock market newsletter that is well written for the most part but includes a section purporting to identify stocks that are likely to rise. It does this by looking for a certain pattern in the stock price. It recently reported, for example, that a certain stock was following the pattern. It also reported that the stock rose six times after the last nine times that pattern occurred. The writers concluded that the probability of the stock rising again was therefore 6/9 or 66.7%.

Using prop.test, we can obtain the confidence interval for the true proportion of times the stock rises after the pattern. Here, the number of observations is *n* = 9 and the number of successes is *x* = 6. The output shows a confidence interval of (0.309, 0.910) at the 95% confidence level:

```
> prop.test(6, 9)

        1-sample proportions test with continuity correction

data:  6 out of 9, null probability 0.5
X-squared = 0.4444, df = 1, p-value = 0.505
alternative hypothesis: true p is not equal to 0.5
95 percent confidence interval:
 0.3091761 0.9095817
sample estimates:
        p
0.6666667
```

The writers are pretty foolish to say the probability of rising is 66.7%. They could be leading their readers into a very bad bet.

By default, prop.test calculates a confidence interval at the 95% confidence level. Use the conf.level argument for other confidence levels:

```
> prop.test(n, x, p, conf.level=0.99)      # 99% confidence level
```

See Also

See Recipe 9.11.

9.13 Testing for Normality

Problem

You want a statistical test to determine whether your data sample is from a normally distributed population.

Solution

Use the shapiro.test function:

```
> shapiro.test(x)
```

The output includes a p-value. Conventionally, $p < 0.05$ indicates that the population is likely not normally distributed whereas $p > 0.05$ provides no such evidence.

Discussion

This example reports a p-value of 0.4151 for x:

```
> shapiro.test(x)

        Shapiro-Wilk normality test

data:  x
W = 0.9651, p-value = 0.4151
```

The large *p*-value suggests the underlying population could be normally distributed. The next example reports a small *p*-value for *y*, so it is unlikely that this sample came from a normal population:

```
> shapiro.test(y)

        Shapiro-Wilk normality test

data:  y
W = 0.9503, p-value = 0.03520
```

I have highlighted the Shapiro–Wilk test because it is a standard R function. You can also install the package `nortest`, which is dedicated entirely to tests for normality. This package includes:

- Anderson–Darling test (`ad.test`)
- Cramer–von Mises test (`cvm.test`)
- Lilliefors test (`lillie.test`)
- Pearson chi-squared test for the composite hypothesis of normality (`pearson.test`)
- Shapiro–Francia test (`sf.test`)

The problem with all these tests is their null hypothesis: they all assume that the population is normally distributed until proven otherwise. As a result, the population must be decidedly nonnormal before the test reports a small *p*-value and you can reject that null hypothesis. That makes the tests quite conservative, tending to err on the side of normality.

Instead of depending solely upon a statistical test, I suggest also using histograms (Recipe 10.18) and quantile-quantile plots (Recipe 10.21) to evaluate the normality of any data. Are the tails too fat? Is the peak to peaked? Your judgment is likely better than a single statistical test.

See Also

See Recipe 3.9 for how to install the `nortest` package.

9.14 Testing for Runs

Problem

Your data is a sequence of binary values: yes–no, 0–1, true–false, or other two-valued data. You want to know: Is the sequence random?

Solution

The `tseries` package contains the `runs.test` function, which checks a sequence for randomness. The sequence should be a factor with two levels:

```
> library(tseries)
> runs.test(as.factor(s))
```

The `runs.test` function reports a *p*-value. Conventionally, a *p*-value of less than 0.05 indicates that the sequence is likely not random whereas a *p*-value exceeding 0.05 provides no such evidence.

Discussion

A *run* is a subsequence composed of identical values, such as all 1s or all 0s. A random sequence should be properly jumbled up, without too many runs. Similarly, it shouldn't contain too *few* runs, either. A sequence of perfectly alternating values (0, 1, 0, 1, 0, 1, ...) contains no runs, but would you say that it's random?

The `runs.test` function checks the number of runs in your sequence. If there are too many or too few, it reports a small *p*-value.

This first example generates a random sequence of 0s and 1s and then tests the sequence for runs. Not surprisingly, `runs.test` reports a large *p*-value, indicating the sequence is likely random:

```
> library(tseries)
> s <- sample(c(0,1), 100, replace=T)
> runs.test(as.factor(s))

        Runs Test

data:  as.factor(s)
Standard Normal = 0.2175, p-value = 0.8279
alternative hypothesis: two.sided
```

This next sequence, however, consists of three runs and so the reported *p*-value is quite low:

```
> s <- c(0,0,0,0,1,1,1,1,0,0,0,0)
> runs.test(as.factor(s))

        Runs Test

data:  as.factor(s)
Standard Normal = -2.2997, p-value = 0.02147
alternative hypothesis: two.sided
```

See Also

See Recipes 5.4 and 8.6.

9.15 Comparing the Means of Two Samples

Problem

You have one sample each from two populations. You want to know if the two populations could have the same mean.

Solution

Perform a *t* test by calling the `t.test` function:

```
> t.test(x, y)
```

By default, `t.test` assumes that your data are not paired. If the observations are paired (i.e., if each x_i is paired with one y_i), then specify `paired=TRUE`:

```
> t.test(x, y, paired=TRUE)
```

In either case, `t.test` will compute a *p*-value. Conventionally, if $p < 0.05$ then the means are likely different whereas $p > 0.05$ provides no such evidence:

- If either sample size is small, then the populations must be normally distributed. Here, "small" means fewer than 20 data points.
- If the two populations have the same variance, specify `var.equal=TRUE` to obtain a less conservative test.

Discussion

I often use the *t* test to get a quick sense of the difference between two population means. It requires that the samples be large enough (both samples have 20 or more observations) or that the underlying populations be normally distributed. I don't take the "normally distributed" part too literally. Being bell-shaped should be good enough.

A key distinction here is whether or not your data contains paired observations, since the results may differ in the two cases. Suppose we want to know if coffee in the morning improves scores on SAT tests. We could run the experiment two ways:

1. Randomly select one group of people. Give them the SAT test twice, once with morning coffee and once without morning coffee. For each person, we will have two SAT scores. These are *paired observations*.
2. Randomly select two groups of people. One group has a cup of morning coffee and takes the SAT test. The other group just takes the test. We have a score for each person, but the scores are not paired in any way.

Statistically, these experiments are quite different. In experiment 1, there are two observations for each person (caffeinated and decaf) and they are not statistically independent. In experiment 2, the data are independent.

If you have paired observations (experiment 1) and erroneously analyze them as unpaired observations (experiment 2), then you could get this result with a *p*-value of 0.9867:

```
> t.test(x,y)

        Welch Two Sample t-test

data:  x and y
t = -0.0166, df = 198, p-value = 0.9867
alternative hypothesis: true difference in means is not equal to 0
95 percent confidence interval:
 -30.56737  30.05605
sample estimates:
mean of x mean of y
 501.2008  501.4565
```

The large *p*-value forces you to conclude there is no difference between the groups. Contrast that result with the one that follows from analyzing the same data but correctly identifying it as paired:

```
> t.test(x, y, paired=TRUE)

        Paired t-test

data:  x and y
t = -2.3636, df = 99, p-value = 0.02005
alternative hypothesis: true difference in means is not equal to 0
95 percent confidence interval:
 -0.4702824 -0.0410375
sample estimates:
mean of the differences
              -0.2556599
```

The *p*-value plummets to 0.02005, and we reach the exactly opposite conclusion.

See Also

If the populations are not normally distributed (bell-shaped) and either sample is small, consider using the Wilcoxon–Mann–Whitney test described in Recipe 9.16.

9.16 Comparing the Locations of Two Samples Nonparametrically

Problem

You have samples from two populations. You don't know the distribution of the populations, but you know they have similar shapes. You want to know: Is one population shifted to the left or right compared with the other?

Solution

You can use a nonparametric test, the Wilcoxon–Mann–Whitney test, which is implemented by the `wilcox.test` function. For paired observations (every x_i is paired with y_i), set `paired=TRUE`:

```
> wilcox.test(x, y, paired=TRUE)
```

For unpaired observations, let `paired` default to `FALSE`:

```
> wilcox.test(x, y)
```

The test output includes a *p*-value. Conventionally, a *p*-value of less than 0.05 indicates that the second population is likely shifted left or right with respect to the first population whereas a *p*-value exceeding 0.05 provides no such evidence.

Discussion

When we stop making assumptions regarding the distributions of populations, we enter the world of *nonparametric statistics*. The Wilcoxon–Mann–Whitney test is nonparametric and so can be applied to more datasets than the *t* test, which requires that the data be normally distributed (for small samples). This test's only assumption is that the two populations have the same shape.

In this recipe, we are asking: Is the second population shifted left or right with respect to the first? This is similar to asking whether the average of the second population is smaller or larger than the first. However, the Wilcoxon–Mann–Whitney test answers a different question: it tells us whether the central locations of the two populations are significantly different or, equivalently, whether their relative frequencies are different.

Suppose we randomly select a group of employees and ask each one to complete the same task under two different circumstances: under favorable conditions and under unfavorable conditions, such as a noisy environment. We measure their completion times under both conditions, so we have two measurements for each employee. We want to know if the two times are significantly different, but we can't assume they are normally distributed.

The data are paired, so we must set `paired=TRUE`:

```
> wilcox.test(fav, unfav, paired=TRUE)

        Wilcoxon signed rank test with continuity correction

data:  fav and unfav
V = 0, p-value < 2.2e-16
alternative hypothesis: true location shift is not equal to 0
```

The *p*-value is essentially zero. Statistically speaking, we reject the assumption that the completion times were equal. Practically speaking, it's reasonable to conclude that the times were different.

In this example, setting `paired=TRUE` is critical. Treating the data as unpaired would be wrong because the observations are not independent; and this, in turn, would produce bogus results. Running the example with `paired=FALSE` produces a *p*-value of 0.2298, which leads to the wrong conclusion.

See Also

See Recipe 9.15 for the parametric test.

9.17 Testing a Correlation for Significance

Problem

You calculated the correlation between two variables, but you don't know if the correlation is statistically significant.

Solution

The `cor.test` function can calculate both the *p*-value and the confidence interval of the correlation. If the variables came from normally distributed populations then use the default measure of correlation, which is the Pearson method:

```
> cor.test(x, y)
```

For nonnormal populations, use the Spearman method instead:

```
> cor.test(x, y, method="Spearman")
```

The function returns several values, including the *p*-value from the test of significance. Conventionally, $p < 0.05$ indicates that the correlation is likely significant whereas $p > 0.05$ indicates it is not.

Discussion

In my experience, people often fail to check a correlation for significance. In fact, many people are unaware that a correlation can be insignificant. They jam their data into a computer, calculate the correlation, and blindly believe the result. However, they should ask themselves: Was there enough data? Is the magnitude of the correlation large enough? Fortunately, the `cor.test` function answers those questions.

Suppose we have two vectors, *x* and *y*, with values from normal populations. We might be very pleased that their correlation is greater than 0.83:

```
> cor(x, y)
[1] 0.8352458
```

But that is naïve. If we run `cor.test`, it reports a relatively large *p*-value of 0.1648:

```
> cor.test(x, y)

        Pearson's product-moment correlation

data:  x and y
t = 2.1481, df = 2, p-value = 0.1648
alternative hypothesis: true correlation is not equal to 0
95 percent confidence interval:
 -0.6379590  0.9964437
sample estimates:
      cor
0.8352458
```

The *p*-value is above the conventional threshold of 0.05, so we conclude that the correlation is unlikely to be significant.

You can also check the correlation by using the confidence interval. In this example, the confidence interval is (–0.638, 0.996). The interval contains zero and so it is possible that the correlation is zero, in which case there would be no correlation. Again, you could not be confident that the reported correlation is significant.

The cor.test output also includes the point estimate reported by cor (at the bottom, labeled "sample estimates"), saving you the additional step of running cor.

By default, cor.test calculates the Pearson correlation, which assumes that the underlying populations are normally distributed. The Spearman method makes no such assumption because it is nonparametric. Use method="Spearman" when working with nonnormal data.

See Also

See Recipe 2.6 for calculating simple correlations.

9.18 Testing Groups for Equal Proportions

Problem

You have samples from two or more groups. The group's elements are binary-valued: either *success* or *failure*. You want to know if the groups have equal proportions of successes.

Solution

Use the prop.test function with two vector arguments:

```
> ns <- c(ns_1, ns_2, ..., ns_N)
> nt <- c(nt_1, nt_2, ..., nt_N)
> prop.test(ns, nt)
```

These are parallel vectors. The first vector, ns, gives the number of successes in each group. The second vector, nt, gives the size of the corresponding group (often called the number of trials).

The output includes a *p*-value. Conventionally, a *p*-value of less than 0.05 indicates that it is likely the groups' proportions are different whereas a *p*-value exceeding 0.05 provides no such evidence.

Discussion

In Recipe 9.11 we tested a proportion based on one sample. Here, we have samples from several groups and want to compare the proportions in the underlying groups.

I recently taught statistics to 38 students and awarded a grade of A to 14 of them. A colleague taught the same class to 40 students and awarded an A to only 10. I wanted to know: Am I fostering grade inflation by awarding significantly more A grades than she did?

I used prop.test. "Success" means awarding an A, so the vector of successes contains two elements: the number awarded by me and the number awarded by my colleague:

```
> successes <- c(14,10)
```

The number of trials is the number of students in the corresponding class:

```
> trials <- c(38,40)
```

The prop.test output yields a *p*-value of 0.3749:

```
> prop.test(successes, trials)

        2-sample test for equality of proportions with continuity correction

data:  successes out of trials
X-squared = 0.7872, df = 1, p-value = 0.3749
alternative hypothesis: two.sided
95 percent confidence interval:
 -0.1110245  0.3478666
sample estimates:
   prop 1    prop 2
0.3684211 0.2500000
```

The relatively large *p*-value means that we cannot reject the null hypothesis: the evidence does not suggest any difference between my grading and hers.

See Also

See Recipe 9.11.

9.19 Performing Pairwise Comparisons Between Group Means

Problem

You have several samples, and you want to perform a pairwise comparison between the sample means. That is, you want to compare the mean of every sample against the mean of every other sample.

Solution

Place all data into one vector and create a parallel factor to identify the groups. Use pairwise.t.test to perform the pairwise comparison of means:

```
> pairwise.t.test(x,f)     # x contains the data, f is the grouping factor
```

The output contains a table of p-values, one for each pair of groups. Conventionally, if $p < 0.05$ then the two groups likely have different means whereas $p > 0.05$ provides no such evidence.

Discussion

This is more complicated than Recipe 9.15, where we compared the means of two samples. Here we have several samples and want to compare the mean of every sample against the mean of every other sample.

Statistically speaking, pairwise comparisons are tricky. It is not the same as simply performing a t test on every possible pair. The p-values must be adjusted, for otherwise you will get an overly optimistic result. The help pages for pairwise.t.test and p.adjust describe the adjustment algorithms available in R. Anyone doing serious pairwise comparisons is urged to review the help pages and consult a good textbook on the subject.

Suppose we are using the data of Recipe 5.5, where we combined data for freshmen, sophomores, and juniors into a data frame called comb. The data frame has two columns: the data in a column called values, and the grouping factor in a column called ind. We can use pairwise.t.test to perform pairwise comparisons between the groups:

```
> pairwise.t.test(comb$values, comb$ind)

        Pairwise comparisons using t tests with pooled SD

data:  comb$values and comb$ind

     fresh  jrs
jrs  0.0043 -
soph 0.0193 0.1621

P value adjustment method: holm
```

Notice the table of *p*-values. The comparisons of juniors versus freshmen and of sophomores versus freshmen produced small *p*-values: 0.0043 and 0.0193, respectively. We can conclude there are significant differences between those groups. However, the comparison of sophomores versus juniors produced a (relatively) large *p*-value of 0.1621, so they are not significantly different.

See Also

See Recipes 5.5 and 9.15.

9.20 Testing Two Samples for the Same Distribution

Problem

You have two samples, and you are wondering: Did they come from the same distribution?

Solution

The Kolmogorov–Smirnov test compares two samples and tests them for being drawn from the same distribution. The ks.test function implements that test:

```
> ks.test(x, y)
```

The output includes a *p*-value. Conventionally, a *p*-value of less than 0.05 indicates that the two samples (*x* and *y*) were drawn from different distributions whereas a *p*-value exceeding 0.05 provides no such evidence.

Discussion

The Kolmogorov–Smirnov test is wonderful for two reasons. First, it is a nonparametric test and so you needn't make any assumptions regarding the underlying distributions: it works for all distributions. Second, it checks the location, dispersion, and shape of the populations, based on the samples. If these characteristics disagree then the test will detect that, allowing us to conclude that the underlying distributions are different.

Suppose we suspect that the vectors *x* and *y* come from differing distributions. Here, ks.test reports a *p*-value of 0.01297:

```
> ks.test(x, y)

        Two-sample Kolmogorov-Smirnov test

data:  x and y
D = 0.3333, p-value = 0.01297
alternative hypothesis: two-sided
```

From the small *p*-value we can conclude that the samples are from different distributions. However, when we test *x* against another sample, *z*, the *p*-value is much larger (0.8245); this suggests that *x* and *z* could have the same underlying distribution:

```
> ks.test(x, z)

        Two-sample Kolmogorov-Smirnov test

data:  x and z
D = 0.1333, p-value = 0.8245
alternative hypothesis: two-sided
```

Graphics

Introduction

Graphics is a great strength of R. The `graphics` package is part of the standard distribution and contains many useful functions for creating a variety of graphic displays. This chapter focuses on those functions, although it occasionally suggests other packages. In this chapter's See Also sections I mention functions in other packages that do the same job in a different way. I suggest that you explore those alternatives if you are dissatisfied with the basic function.

Graphics is a vast subject, and I can only scratch the surface here. If you want to delve deeper, I recommend *R Graphics* by Paul Murrell (Chapman & Hall, 2006). That book discusses the paradigms behind R graphics, explains how to use the graphics functions, and contains numerous examples—including the code to recreate them. Some of the examples are pretty amazing.

The Illustrations

The graphs in this chapter are mostly plain and unadorned. I did that intentionally. When you call the `plot` function, as in:

```
> plot(x)
```

you get a plain, graphical representation of *x*. You could adorn the graph with colors, a title, labels, a legend, text, and so forth, but then the call to `plot` becomes more and more crowded, obscuring the basic intention:

```
> plot(x, main="Forecast Results", xlab="Month", ylab="Production",
+     col=c("red", "black", "green"))
```

I want to keep the recipes clean, so I emphasize the basic plot and then show later (as in Recipe 10.2) how to add adornments.

Notes on Graphics Functions

It is important to understand the distinction between high-level and low-level graphics functions. A *high-level graphics function* starts a new graph. It initializes the graphics window (creating it if necessary); sets the scale; maybe draws some adornments, such as a title and labels; and renders the graphic. Examples include:

plot
> Generic plotting function

boxplot
> Create a box plot

hist
> Create a histogram

qqnorm
> Create a quantile-quantile (Q-Q) plot

curve
> Graph a function

A *low-level graphics function* cannot start a new graph. Rather, it adds something to an existing graph: points, lines, text, adornments, and so forth. Examples include:

points
> Add points

lines
> Add lines

abline
> Add a straight line

segments
> Add line segments

polygon
> Add a closed polygon

text
> Add text

You must call a high-level graphics routine *before* calling a low-level graphics routine. The low-level routine needs to have the graph initialized; otherwise, you get an error like this:

```
> abline(a=0, b=1)
Error in int_abline(a = a, b = b, h = h, v = v, untf = untf, ...) :
  plot.new has not been called yet
```

The Generic plot Function

In this chapter, we come face to face with polymorphism in R. A *polymorphic function* or *generic function* is one whose behavior changes depending on the type of argument. The `plot` function is polymorphic, so `plot(x)` produces different results depending on whether *x* is a vector, a factor, a data frame, a linear regression model, a table, or whatever.

As you read the recipes, you'll see `plot` used over and over. Each time it's used, carefully note the type of arguments. That will help you understand and remember why the function behaved as it did.

Graphics in Other Packages

R is highly programmable, and many people have extended its graphics machinery with additional features. Quite often, packages include specialized functions for plotting their results and objects. The `zoo` package, for example, implements a time series object. If you create a `zoo` object `z` and call `plot(z)`, then the `zoo` package does the plotting; it creates a graphic that is customized for displaying a time series.

There are even entire packages devoted to extending R with new graphics paradigms. The `lattice` package is an alternative to traditional graphics. It uses a more powerful graphics paradigm that enables you to create informative graphics more easily. The results are generally better looking, too. It was implemented by Deepayan Sarkar, who also wrote *Lattice: Multivariate Data Visualization with R* (Springer, 2008), which explains the package and how to use it. The lattice package is also described in *R in a Nutshell* (O'Reilly).

The `ggplot2` package provides yet another graphics paradigm, which is called the Grammar of Graphics. It uses a highly modular approach to graphics, which lets you construct and customize your plots more easily. These graphics, too, are generally more attractive than the traditional ones. The `ggplot2` package was created by Hadley Wickham, who also wrote *ggplot2: Elegant Graphics for Data Analysis* (Springer, 2009). That book explains the paradigm behind `ggplot2` and how to use the package functions.

10.1 Creating a Scatter Plot

Problem

You have paired observations: $(x_1, y_1), (x_2, y_2), ..., (x_n, y_n)$. You want to create a scatter plot of the pairs.

Solution

If your data are held in two parallel vectors, *x* and *y*, then use them as arguments of `plot`:

```
> plot(x, y)
```

If your data is held in a (two-column) data frame, plot the data frame:

```
> plot(dfrm)
```

Discussion

A scatter plot is usually my first attack on a new dataset. It's a quick way to see the relationship, if any, between *x* and *y*. Creating the scatter plot is easy:

```
> plot(x, y)
```

The plot function does not return anything. Rather, its purpose is to draw a plot of the (*x*, *y*) pairs in the graphics window.

Life is even easier if your data is captured in a two-column data frame. If you plot a two-column data frame, the function assumes you want a scatter plot created from the two columns. The scatter plot shown in Figure 10-1 was created by one call to plot:

```
> plot(cars)
```

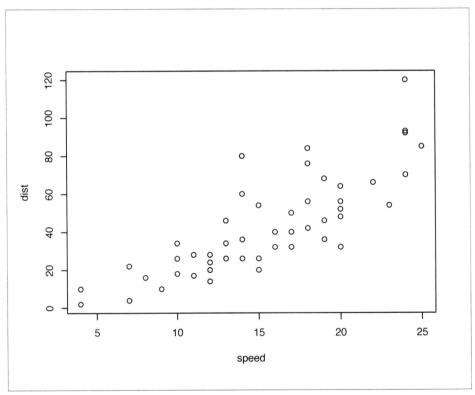

Figure 10-1. Scatter plot

The cars dataset contains two columns, speed and dist. The first column is speed, so that becomes the *x*-axis and dist becomes the *y*-axis.

If your data frame contains more than two columns then you will get multiple scatter plots, which might or might not be useful (Recipe 10.7).

To get a scatter plot, your data must be numeric. Recall that `plot` is a polymorphic function and so, if the arguments are nonnumeric, it will create some other plot. See Recipe 10.16, for example, which creates box plots from factors.

See Also

See Recipe 10.2 for adding a title and labels; see Recipes 10.3 and 10.5 for adding a grid and a legend (respectively). See Recipe 10.7 for plotting multiple variables.

10.2 Adding a Title and Labels

Problem

You want to add a title to your plot or add labels for the axes.

Solution

When calling `plot`:

- Use the `main` argument for a title.
- Use the `xlab` argument for an x-axis label.
- Use the `ylab` argument for a y-axis label.

```
> plot(x, main="The Title", xlab="X-axis Label", ylab="Y-axis Label")
```

Alternatively: plot your data but set `ann=FALSE` to inhibit annotations; then call the `title` function to add a title and labels:

```
> plot(x, ann=FALSE)
> title(main="The Title", xlab="X Axis Label", ylab="Y Axis Label")
```

Discussion

The graphic in Recipe 10.1 is quite plain. It needs a title and better labels to make it more interesting and easier to interpret. We can add them via arguments to `plot`, yielding the graphic shown in Figure 10-2:

```
> plot(cars,
+     main="cars: Speed vs. Stopping Distance (1920)",
+     xlab="Speed (MPH)",
+     ylab="Stopping Distance (ft)")
```

This is a clear improvement over the defaults. Now the title tells us something about the plot, and the labels give the measurement units.

Figure 10-2. Adding a title and labels

10.3 Adding a Grid

Problem

You want to add a grid to your graphic.

Solution

- Call plot with type="n" to initialize the graphics frame without displaying the data.
- Call the grid function to draw the grid.
- Call low-level graphics functions, such as points and lines, to draw the graphics overlaid on the grid.

```
> plot(x, y, type="n")
> grid()
> points(x, y)
```

Discussion

The plot function does not automatically draw a grid. For some plots, however, a grid is helpful to the viewer.

The Solution shows the proper way to add a grid. The grid must be drawn first, before the graphics. If you draw the graphics first, the grid will overlay them and partially hide them.

We can add a grid to the graphic of Recipe 10.2. First, we draw the graphic but with type="n". That will create a new plot, including title and axes, but will not (yet) plot the data:

```
> plot(cars,
+       main="cars: Speed vs. Stopping Distance (1920)",
+       xlab="Speed (MPH)",
+       ylab="Stopping Distance (ft)",
+       type="n")
```

Next we draw the grid:

```
> grid()
```

Finally, we draw the scatter plot by calling points, a low-level graphics function, thus creating the graph shown in Figure 10-3:

```
> points(cars)
```

In a hurry? Looking for a shortcut? You can call plot as you normally would and then call grid. The only problem is that the grid will be overlaid on your graphics, partially overwriting them. Sometimes this creates a visible problem and sometimes not. Try it.

10.4 Creating a Scatter Plot of Multiple Groups

Problem

You have paired observations in two vectors, x and y, and a parallel factor f that indicates their groups. You want to create a scatter plot of x and y that distinguishes among the groups.

Solution

Use the pch argument of plot. It will plot each point with a different plotting character, according to its group:

```
> plot(x, y, pch=as.integer(f))
```

Figure 10-3. Adding a grid

Discussion

Plotting multiple groups in one scatter plot creates an uninformative mess unless we distinguish one group from another. The pch argument of plot lets us plot points using a different plotting character for each (x, y) pair.

The iris dataset contains paired measures of Petal.Length and Petal.Width. Each measurement also has a Species property indicating the species of the flower that was measured. If we plot all the data at once, we just get the scatter plot shown in the lefthand panel of Figure 10-4:

```
> with(iris, plot(Petal.Length, Petal.Width))
```

The graphic would be far more informative if we distinguished the points by species. The pch argument is a vector of integers, one integer for each (x, y) point, with values between 0 and 18. The point will be drawn according to its corresponding value in pch. This call to plot draws each point according to its group:

```
> with(iris, plot(Petal.Length, Petal.Width, pch=as.integer(Species)))
```

The result is displayed in the righthand panel of Figure 10-4, which clearly shows the points in groups according to their species.

Figure 10-4. Scatter plots of multiple groups

See Also

See Recipe 10.5 to add a legend.

10.5 Adding a Legend

Problem

You want your plot to include a *legend*, the little box that decodes the graphic for the viewer.

Solution

After calling plot, call the legend function. The first three arguments are always the same, but the arguments after that vary as a function of what you are labeling. Here is how you create legends for points, lines, and colors:

Legend for points

```
legend(x, y, labels, pch=c(pointtype1, pointtype2, ...))
```

Legend for lines according to line type

```
legend(x, y, labels, lty=c(linetype1, linetype2, ...))
```

Legend for lines according to line width

```
legend(x, y, labels, lwd=c(width1, width2, ...))
```

Legend for colors
```
legend(x, y, labels, col=c(color1, color2, ...))
```

Here, x and y are the coordinates of the legend box, and labels is a vector of the character strings to appear in the legend. The pch, lty, lwd, and col arguments are vectors that parallel the labels. Each label will be shown next to its corresponding point type, line type, line width, or color according to which one you specified.

Discussion

The lefthand panel of Figure 10-5 shows a legend added to the scatter plot of Recipe 10.4. The legend tells the viewer which points are associated with which species. I could have added the legend in this way (after calling plot):

```
> legend(1.5, 2.4, c("setosa","versicolor","virginica"), pch=1:3)
```

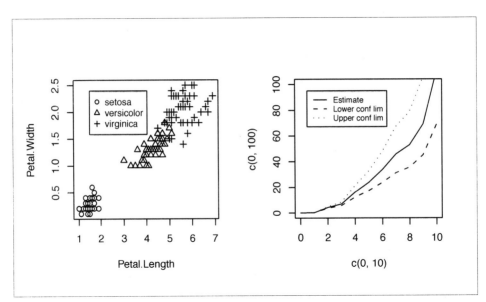

Figure 10-5. Examples of legends

The first two arguments give the x and y coordinates of the legend's box. The third argument is a vector of labels; in this case, names of species. The last argument indicates that we are labeling points by their type—specifically, types 1 through 3.

In fact, I created the plot and legend this way:

```
> f <- factor(iris$Species)
> with(iris, plot(Petal.Length, Petal.Width, pch=as.integer(f)))
> legend(1.5, 2.4, as.character(levels(f)), pch=1:length(levels(f)))
```

The beauty here is that species names are not hard-coded into the R code. Rather, everything is driven by the species factor, f. I like that for two reasons. First, it means that I won't forget or misspell a species name. Second, if someone adds another species then the code automatically adapts to the addition. That second virtue is especially valuable inside an R script.

The righthand panel of Figure 10-5 shows a legend for lines based on the line type (solid, dashed, or dotted). It was created like this:

```
> legend(0.5, 95, c("Estimate","Lower conf lim","Upper conf lim"),
+        lty=c("solid","dashed","dotted"))
```

The two examples in this recipe show legends for point types and line types. The legend function can also create legends for colors, fills, and line widths. The function also has a bewildering number of arguments and options for controlling the exact appearance of the legend. See the help page for details.

Legends are a wonderful feature in general, but the legend function comes with two annoyances. First, you must pick the legend's x and y coordinates yourself. R cannot pick them for you. Plot the data without a legend; then pick a nice, empty area and put the legend there. Otherwise, the legend might obscure the graphics.

Second, the correspondence between the labels and the point types (or line types or colors) is not automatic. You are responsible for getting that right. The task is prone to errors and is not easy to automate.

10.6 Plotting the Regression Line of a Scatter Plot

Problem

You are plotting pairs of data points, and you want to add a line that illustrates their linear regression.

Solution

Create a model object, plot the (x, y) pairs, and then plot the model object using the abline function:

```
> m <- lm(y ~ x)
> plot(y ~ x)
> abline(m)
```

Discussion

When you plot a model object using abline(m), it draws the fitted regression line. This is very different from plot(m), which produces diagnostic plots (Recipe 11.15).

Suppose we are modeling the `strongx` dataset found in the `faraway` package:

```
> library(faraway)
> data(strongx)
> m <- lm(crossx ~ energy, data=strongx)
```

We can display the data in a scatter plot and then add the regression line using `abline`, producing Figure 10-6:

```
> plot(crossx ~ energy, data=strongx)
> abline(m)
```

Notice that we use a formula to define the plot (`crossx ~ energy`). I like using that syntax because we quickly see the parallel with the regression formula. The alternative is calling `plot(strongx$energy, strongx$crossx)`, which trips me up because the order of arguments is reversed from the `lm` formula.

See Also

See Chapter 11 for more about linear regression and the `lm` function.

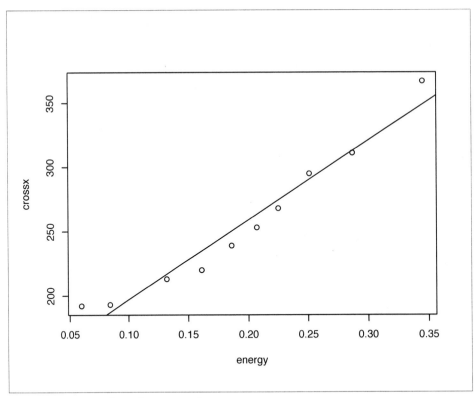

Figure 10-6. Scatter plot with regression line

10.7 Plotting All Variables Against All Other Variables

Problem

Your dataset contains multiple numeric variables. You want to see scatter plots for all pairs of variables.

Solution

Place your data in a data frame and then plot the data frame. R will create one scatter plot for every pair of columns:

```
> plot(dfrm)
```

Discussion

When you have a large number of variables, finding interrelationships between them is difficult. One useful technique is looking at scatter plots of all pairs of variables. This would be quite tedious if done pair-by-pair, but R provides an easy way to produce all those scatter plots at once.

The iris dataset contains four numeric variables and one categorical variable:

```
> head(iris)
  Sepal.Length Sepal.Width Petal.Length Petal.Width Species
1          5.1         3.5          1.4         0.2  setosa
2          4.9         3.0          1.4         0.2  setosa
3          4.7         3.2          1.3         0.2  setosa
4          4.6         3.1          1.5         0.2  setosa
5          5.0         3.6          1.4         0.2  setosa
6          5.4         3.9          1.7         0.4  setosa
```

What is the relationship, if any, between the numeric columns? Plotting those four columns produces the multiple scatter plots shown in Figure 10-7:

```
> plot(iris[,1:4])
```

10.8 Creating One Scatter Plot for Each Factor Level

Problem

Your dataset contains (at least) two numeric variables and a factor. You want to create several scatter plots for the numeric variables, with one scatter plot for each level of the factor.

Solution

This kind of plot is called a *conditioning plot*, which is produced by the coplot function:

```
> coplot(y ~ x | f)
```

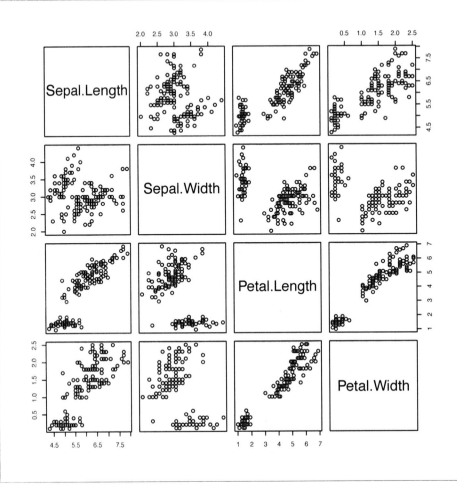

Figure 10-7. Multiple scatter plots

This will produce scatter plots of *x* versus *y*, with one plot for every level of *f*.

Discussion

Conditioning plots are another way to explore and illustrate the effect of a factor on your data.

The Cars93 dataset contains 27 variables describing 93 car models as of 1993. Two numeric variables are MPG.city, the miles per gallon in the city, and Horsepower, the engine horsepower. One categorical variable is Origin, which can be USA or non-USA according to where the model was built.

Exploring the relationship between MPG and horsepower, we might ask: Is there a different relationship for USA models and non-USA models? The conditioning plot in Figure 10-8 shows two scatter plots of MPG versus horsepower that were created simultaneously by one call to coplot. The lefthand plot is for USA vehicles and the righthand plot is for non-USA vehicles. It was produced like this:

```
> data(Cars93, package="MASS")
> coplot(Horsepower ~ MPG.city | Origin, data=Cars93)
```

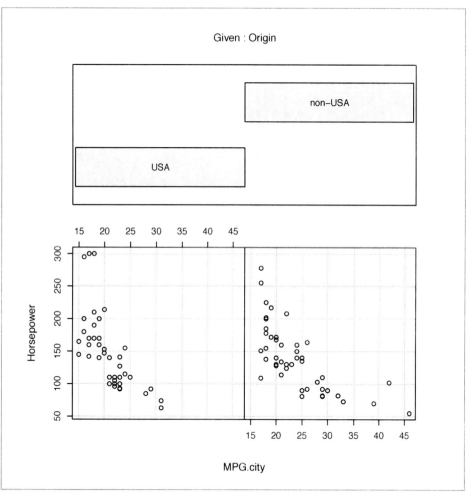

Figure 10-8. A conditioning plot

The data argument of coplot lets us take variables from the Cars93 data frame.

The plot reveals a few insights. If we really crave that 300-horsepower monster then we'll have to buy a car built in the USA; but if we want high MPG, we have more choices among non-USA models.

These insights could be teased out of a statistical analysis, but the visual presentation reveals them much more quickly.

10.9 Creating a Bar Chart

Problem

You want to create a bar chart.

Solution

Use the barplot function. The first argument is a vector of bar heights:

```
> barplot(c(height₁, height₂, ..., heightₙ))
```

Discussion

The barplot function produces a simple bar chart. It assumes that the heights of your bars are conveniently stored in a vector. That is not always the case, however. Often you have a vector of numeric data and a parallel factor that groups the data, and you want to produce a bar chart of the group means or the group totals. For example, the airquality dataset contains a numeric Temp column and a Month column. We can create a bar chart of the mean temperature by month in two steps. First, we compute the means:

```
> heights <- tapply(airquality$Temp, airquality$Month, mean)
```

That gives the heights of the bars, from which we create the bar chart:

```
> barplot(heights)
```

The result is shown in the lefthand panel of Figure 10-9. The result is pretty bland, as you can see, so it's common to add some simple adornments: a title, labels for the bars, and a label for the y-axis:

```
> barplot(heights,
+         main="Mean Temp. by Month",
+         names.arg=c("May", "Jun", "Jul", "Aug", "Sep"),
+         ylab="Temp (deg. F)")
```

The righthand panel of Figure 10-9 shows the improved bar chart.

See Also

See Recipe 10.10 for adding confidence intervals and Recipe 10.11 for adding color; see the barchart function in the lattice package.

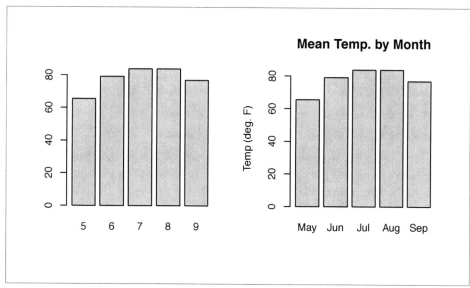

Figure 10-9. Bar charts, plain and adorned

10.10 Adding Confidence Intervals to a Bar Chart

Problem

You want to augment a bar chart with confidence intervals.

Solution

Suppose that x is a vector of statistics, such as a vector of means, and that lower and upper are vectors of the corresponding limits for the confidence intervals. The barplot2 function of the gplots library can display a bar chart of x and its confidence intervals:

```
> library(gplots)
> barplot2(x, plot.ci=TRUE, ci.l=lower, ci.u=upper)
```

Discussion

Most bar charts display point estimates, which are shown by the heights of the bars, but rarely do they include confidence intervals. The statistician in me dislikes this intensely. The point estimate is only half of the story; the confidence interval gives the full story.

Fortunately, the barplot2 function can plot a bar chart with confidence intervals. The hard part is calculating the intervals. In Recipe 10.9, we calculated group means before plotting them. We can calculate the confidence intervals for those means in a similar way. The t.test function returns a list of values. That list includes an element called

conf.int, which is itself a vector of two elements: the lower and upper bounds of the confidence interval for the mean:

```
> attach(airquality)
> heights <- tapply(Temp, Month, mean)
> lower <- tapply(Temp, Month, function(v) t.test(v)$conf.int[1])
> upper <- tapply(Temp, Month, function(v) t.test(v)$conf.int[2])
```

We could create a basic bar chart with confidence intervals in this way:

```
> barplot2(heights, plot.ci=TRUE, ci.l=lower, ci.u=upper)
```

But as noted in Recipe 10.9, that would create a very bland graphic. The bar chart is more attractive if we limit the y range (ylim), trim the bars (xpd), add a title (main), label the bars (names.arg), and label the y-axis. The bar chart in Figure 10-10 was produced by this call to barplot2:

```
> barplot2(heights, plot.ci=TRUE, ci.l=lower, ci.u=upper,
+          ylim=c(50,90), xpd=FALSE,
+          main="Mean Temp. By Month",
+          names.arg=c("May","Jun","Jul","Aug","Sep"),
+          ylab="Temp (deg. F)")
```

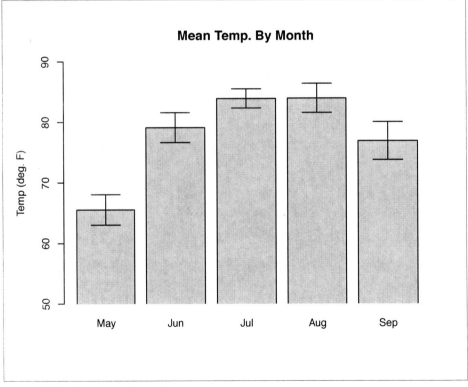

Figure 10-10. Bar chart with confidence intervals

See Also

See Recipe 3.9 to install the `gplots` package, Recipe 6.5 for more about `tapply`, and Recipe 9.9 for more about `t.test`.

10.11 Coloring a Bar Chart

Problem

You want to color or shade the bars of a bar chart.

Solution

Use the `col` argument of barplot:

```
> barplot(heights, col=colors)
```

Here, `heights` is the vector of bar heights and `colors` is a vector of corresponding colors.

To generate the vector of colors, you would typically use the `gray` function to generate a vector of grays or the `rainbow` function to generate a vector of colors.

Discussion

Building a vector of colors can be tricky. Using explicit colors is a simple solution. This little example plots a three-bar chart and colors the bars red, white, and blue (respectively):

```
> barplot(c(3,5,4), col=c("red","white","blue"))
```

Life is rarely that simple, however. More likely, you want colors that convey some information about your dataset. A typical effect is to shade the bars according to their rank: shorter bars are light colored; taller bars are darker. In this example, I'll use gray-scale colors generated by the `gray` function. The one argument of `gray` is a vector of numeric values between 0 and 1. The function returns one shade of gray for each vector element, ranging from pure black for 0.0 to pure white for 1.0.

To shade the bar chart, we first convert the bars' ranks to relative heights, expressed as a value between zero and one:

```
> rel.hts <- rank(heights) / length(heights)
```

Then we convert the relative heights into a vector of grayscale colors while inverting the relative heights so that the taller bars are dark, not light:

```
> grays <- gray(1 - rel.hts)
```

We could easily create a shaded bar chart in this way:

```
> barplot(heights, col=grays)
```

However, we'll add adornments to make the chart easier to interpret. Here is the complete solution, with the result shown in Figure 10-11:

```
> rel.hts <- (heights - min(heights)) / (max(heights) - min(heights))
> grays <- gray(1 - rel.hts)
> barplot(heights,
+         col=grays,
+         ylim=c(50,90), xpd=FALSE,
+         main="Mean Temp. By Month",
+         names.arg=c("May", "Jun", "Jul", "Aug", "Sep"),
+         ylab="Temp (deg. F)")
```

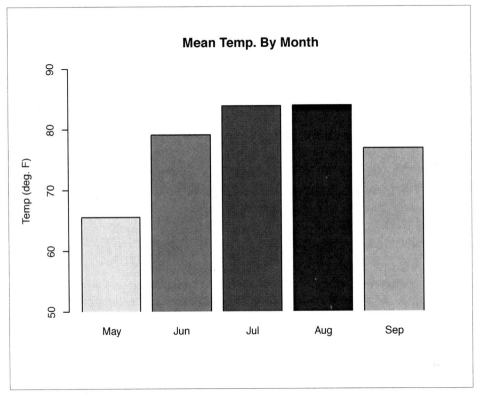

Figure 10-11. Bar chart with shading

I used grayscale here because some editions of this book are not printed in color. For a multichromatic effect I suggest using the rainbow function, which can produce a range of colors; use it instead of the gray function used here.

See Also

See Recipe 10.9 for creating a bar chart.

10.12 Plotting a Line from x and y Points

Problem

You have paired observations: (x_1, y_1), (x_2, y_2), ..., (x_n, y_n). You want to plot a series of line segments that connect the data points.

Solution

Use the plot function with a plot type of "l" (the letter ell, not the digit one):

```
> plot(x, y, type="l")
```

If your data is captured in a two-column data frame, you can plot the line segments from data frame contents:

```
> plot(dfrm, type="l")
```

Discussion

Observe that this recipe makes the same assumption as Recipe 10.1 for creating a scatter plot—you have paired observations. The only difference between a scatter plot and a line drawing is that one plots points and the other connects them. The plot function handles both cases. The type argument determines scatter plot or line drawing; the default is a scatter plot.

We can plot a built-in dataset, pressure, both ways: as a scatter plot and a line plot. The left panel of Figure 10-12 show the scatter plot, created as follows:

```
> plot(pressure)
```

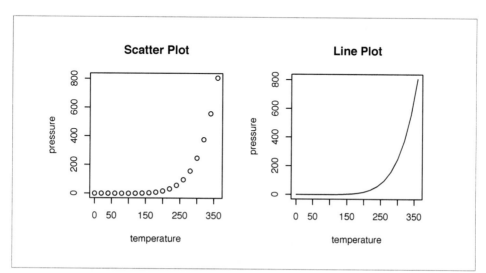

Figure 10-12. Same dataset, two plot types

The right panel is a line plot, which was created simply by changing the plot type to "line" (1):

```
> plot(pressure, type="l")
```

See Also

See Recipe 10.1.

10.13 Changing the Type, Width, or Color of a Line

Problem

You are plotting a line. You want to change the type, width, or color of the line.

Solution

The plot function (and other graphics functions) include parameters for controlling the appearance of lines. Use the lty parameter to control the line type:

- lty="solid" or lty=1 (default)
- lty="dashed" or lty=2
- lty="dotted" or lty=3
- lty="dotdash" or lty=4
- lty="longdash" or lty=5
- lty="twodash" or lty=6
- lty="blank" or lty=0 (inhibits drawing)

This call to plot, for example, draws lines with a dashed style:

```
> plot(x, y, type="l", lty="dashed")
```

Use the lwd parameter to control the line width or thickness. By default, lines have a width of 1:

```
> plot(x, y, type="l", lwd=2)              # Draw a thicker line
```

Use the col parameter to control line color. By default, lines are drawn in black:

```
> plot(x, y, type="l", col="red")          # Draw a red line
```

Discussion

The Solution shows how to draw one line and specify its style, width, or color. A common scenario involves drawing multiple lines, each with its own style, width, or color. The Solution generalizes to that case, but call the plot function for the first line and call the lines function for subsequent lines. The lines function adds lines to an existing plot and lets you specify their style, width, and/or color:

```
> plot(x, y.democr, type="l", col="blue")
> lines(x, y.republ, col="red")
> lines(x, y.indeps, col="yellow")
```

Here, the initial call to `plot` initializes the graphics window and draws the first lines in blue. The subsequent calls to `lines` draw additional lines, first in red and then in yellow.

See Also

See Recipe 10.12 for plotting a basic line.

10.14 Plotting Multiple Datasets

Problem

You want to show multiple datasets in one plot.

Solution

Initialize the plot using a high-level graphics function such as `plot` or `curve`. Then add additional datasets using low-level functions such as `lines` and `points`.

When initializing the plot, be careful to size the canvas correctly. Using the `xlim` and `ylim` parameters, create a canvas large enough to hold all datasets and not merely the first dataset. This code snippet shows how to compute the limits using the `range` function and then set them using `xlim` and `ylim`. It assumes you have two datasets, one represented by `x1` and `y1` and the other by `x2` and `y2`:

```
> xlim <- range(c(x1,x2))
> ylim <- range(c(y1,y2))
> plot(x1, y1, type="l", xlim=xlim, ylim=ylim)
> lines(x2, y2, lty="dashed")
```

Discussion

As discussed in the "Introduction" to this chapter, the `plot` function initializes the graphics window and the `lines` function adds lines after that. My first try at using plot and lines together looked something like this (I wanted to plot the (`x1`, `y1`) dataset and the (`x2`, `y2`) dataset together):

```
> plot(x1, y1, type="l")
> lines(x2, y2, lty="dashed")
```

The result was something like the lefthand panel of Figure 10-13. The first dataset plotted correctly, but the second dataset just sort of trailed off the page. This behavior was puzzling.

The problem was that the plot function sized the canvas based on the data it was given. The second dataset (plotted by `lines`) had a different, wider range, so it fell outside the canvas and disappeared.

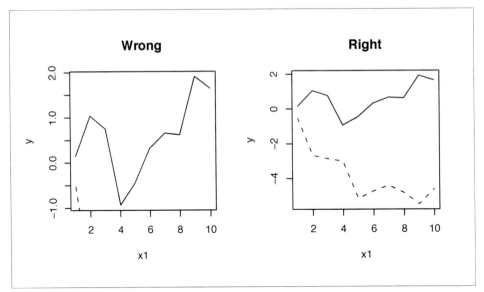

Figure 10-13. Plotting multiple datasets, correctly and incorrectly

The solution is to size the canvas correctly in the first place. The plot function has two arguments for sizing the canvas—xlim is a two-element vector giving the lower and upper limits of the x-axis, and ylim does the same for the y-axis:

```
> plot(x, y, xlim=c(lower,upper), ylim=c(lower,upper), ...)
```

The range function is handy here. It returns the upper and lower bounds of a vector; hence we can use it to compute xlim and ylim before calling plot, like this:

```
> xlim <- range(c(x1,x2))
> ylim <- range(c(y1,y2))
> plot(x1, y1, type="l", xlim=xlim, ylim=ylim)
> lines(x2, y2, lty="dashed")
```

The righthand panel of Figure 10-13 shows the result. Now the canvas is big enough, and no data is lost.

This Discussion has been framed in terms of lines, but plotting points presents the same problem and has the same solution.

See Also

See this chapter's "Introduction" for comments about high- and low-level graphics functions.

10.15 Adding Vertical or Horizontal Lines

Problem

You want to add a vertical or horizontal line to your plot, such as an axis through the origin.

Solution

The `abline` function will draw a vertical line or horizontal line when given an argument of v or h, respectively:

```
> abline(v=x)      # Draw a vertical line at x
> abline(h=y)      # Draw a horizontal line at y
```

Discussion

The `abline` function draws straight lines. I often use it to draw axes through the origin of my plot:

```
> abline(v=0)      # Vertical line at x = 0
> abline(h=0)      # Horizontal line at y = 0
```

The `abline` function adds lines to an existing plot; it cannot create a new plot. So call a plot creation function first, such as `plot(...)`, and then call `abline`.

The v and h arguments can be vectors, in which case `abline` draws multiple lines—one for each vector element. A typical use would be drawing regularly spaced lines. Suppose we have a sample of points, `samp`. First, we plot them with a solid line through the mean:

```
> plot(samp)
> m <- mean(samp)
> abline(h=m)
```

We want to illustrate the sample standard deviation graphically, so we calculate it and draw dotted lines at ±1 and ±2 standard deviations away from the mean:

```
> stdevs <- m + c(-2,-1,+1,+2)*sd(samp)
> abline(h=stdevs, lty="dotted")
```

The result is shown in Figure 10-14.

See Also

See Recipe 10.13 for more about changing line types.

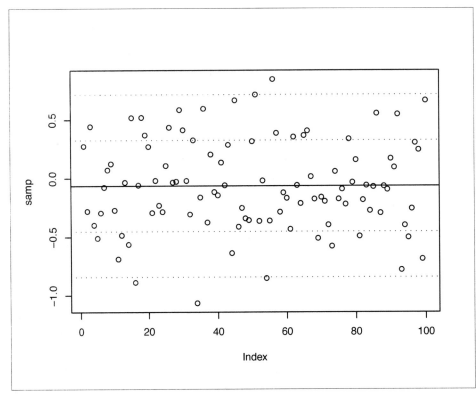

Figure 10-14. Delineating sample standard deviations

10.16 Creating a Box Plot

Problem

You want to create a box plot of your data.

Solution

Use `boxplot(x)`, where x is a vector of numeric values.

Discussion

A box plot provides a quick and easy visual summary of a dataset. Figure 10-15 shows a typical box plot:

- The thick line in the middle is the median.
- The box surrounding the median identifies the first and third quartiles; the bottom of the box is Q1, and the top is Q3.

- The "whiskers" above and below the box show the range of the data, excluding outliers.
- The circles identify outliers. By default, an outlier is defined as any value that is farther than 1.5 × IQR away from the box. (IQR is the *interquartile range*, or Q3 – Q1.) In this example, there are three outliers.

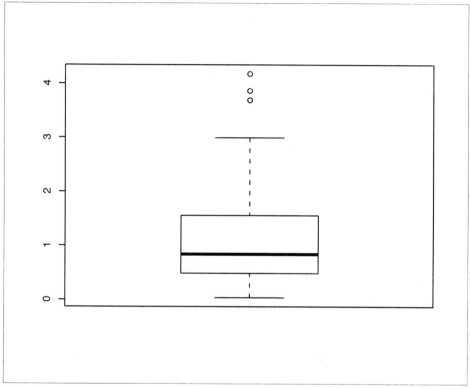

Figure 10-15. A typical box plot

See Also

One box plot alone is pretty boring. See Recipe 10.17 for creating multiple box plots.

10.17 Creating One Box Plot for Each Factor Level

Problem

Your dataset contains a numeric variable and a factor (categorical variable). You want to create several box plots of the numeric variable broken out by factor levels.

Solution

Use the `boxplot` function with a formula:

```
> boxplot(x ~ f)
```

Here, `x` is the numeric variable and `f` is the factor.

You can also use the two-argument form of the `plot` function, being careful to put the factor first:

```
> plot(f, x)
```

Discussion

This recipe is another great way to explore and illustrate the relationship between two variables. In this case, we want to know whether the numeric variable changes according to the level of the factor.

The `UScereal` dataset contains many variables regarding breakfast cereals. One variable is the amount of sugar per portion and another is the shelf position (counting from the floor). Cereal manufacturers can negotiate for shelf position, placing their product for the best sales potential. I wonder: Where do they put the high-sugar cereals? We can explore that question by creating one box plot per shelf, like this:

```
> data(UScereal, package="MASS")
> boxplot(sugars ~ shelf, data=UScereal)
```

The resulting box plot would be quite drab, however, so we add a few adornments:

```
> data(UScereal, package="MASS")
> boxplot(sugars ~ shelf, data=UScereal,
+         main="Sugar Content by Shelf",
+         xlab="Shelf", ylab="Sugar (grams per portion)")
```

The result is shown in Figure 10-16. The `data` argument lets us take variables from the `UScereal` data frame. I supplied `xlab` (the x-axis label) and `ylab` (the y-axis label) because the basic box plot has no labels, making it harder to interpret.

The box plots suggest that shelf #2 has the most high-sugar cereals. Could it be that this shelf is at eye level for young children who can influence their parent's choice of cereals?

See Also

See Recipe 10.16 for creating a basic box plot.

10.18 Creating a Histogram

Problem

You want to create a histogram of your data.

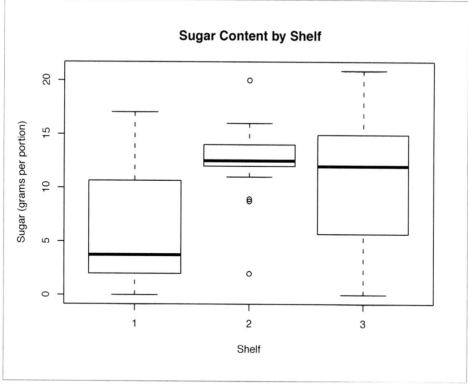

Figure 10-16. Box plots by factor level

Solution

Use hist(x), where x is a vector of numeric values.

Discussion

The lefthand panel of Figure 10-17 shows a histogram of the MPG.city column taken from the Cars93 dataset, created like this:

```
> data(Cars93, package="MASS")
> hist(Cars93$MPG.city)
```

The hist function must decide how many cells (bins) to create for binning the data. In this example, the default algorithm chose seven bins. That creates too few bars for my taste because the shape of the distribution remains hidden. So I would include a second argument for hist—namely, the suggested number of bins:

```
> hist(Cars93$MPG.city, 20)
```

The number is only a suggestion, but hist will expand the number of bins as possible to accommodate that suggestion.

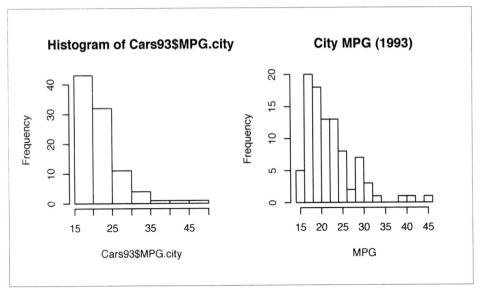

Figure 10-17. Histograms

The righthand panel of Figure 10-17 shows a histogram for the same data but with more bins and with replacements for the default title and x-axis label. It was created like this:

```
> hist(Cars93$MPG.city, 20, main="City MPG (1993)", xlab="MPG")
```

See Also

The histogram function of the lattice package is an alternative to hist.

10.19 Adding a Density Estimate to a Histogram

Problem

You have a histogram of your data sample, and you want to add a curve to illustrate the apparent density.

Solution

Use the density function to approximate the sample density; then use lines to draw the approximation:

```
> hist(x, prob=T)            # Histogram of x, using a probability scale
> lines(density(x))          # Graph the approximate density
```

Discussion

A histogram suggests the density function of your data, but it is rough. A smoother estimate could help you better visualize the underlying distribution.

This example takes a sample from a gamma distribution and then plots the histogram and the estimated density. The result is shown in Figure 10-18:

```
> samp <- rgamma(500, 2, 2)
> hist(samp, 20, prob=T)
> lines(density(samp))
```

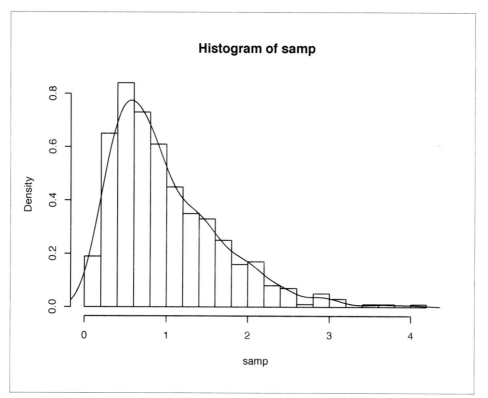

Figure 10-18. Histogram with density estimate

You must include the prob=T argument to hist in order for this recipe to work. The result of prob=T is that the y-axis of Figure 10-18 is given in probabilities. Without it, hist gives counts instead; that would be incompatible with plotting the output of density.

See Also

The density function approximates the shape of the density nonparametrically. If you know the actual underlying distribution, use instead Recipe 8.11 to plot the density function.

10.20 Creating a Discrete Histogram

Problem

You want to create a histogram of discrete data.

Solution

Use the `table` function to count occurrences. Then use the `plot` function with `type="h"` to graph the occurrances as a histogram:

```
> plot(table(x), type="h")
```

Here, x is a vector of discrete values.

Discussion

You can use the `hist` function to create a histogram of discrete data, but `hist` is really oriented toward continuous data. The resulting histograms don't have that look of "discreteness".

Figure 10-19 shows a histogram created using this function call:

```
> plot(table(x), type="h", lwd=5, ylab="Freq")
```

The `lwd=5` parameter widens the bars a bit, giving them a more pleasing appearance.

If you prefer a histogram with relative frequencies, not counts, then simply scale the table elements by the number of data items, like this:

```
> plot(table(x)/length(x), type="h", lwd=5, ylab="Freq")
```

See Also

See Recipe 10.18 for histograms of continuous data. This recipe was taken straight from the help page for `plot`.

10.21 Creating a Normal Quantile-Quantile (Q-Q) Plot

Problem

You want to create a quantile-quantile (Q-Q) plot of your data, typically because you want to know whether the data is normally distributed.

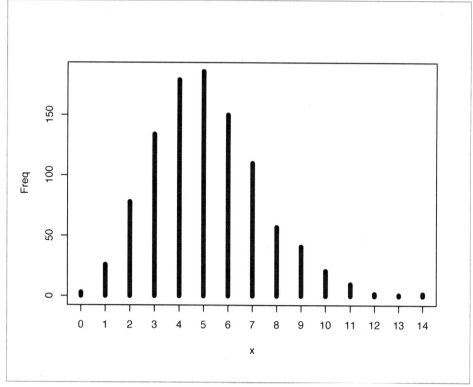

Figure 10-19. Histogram of discrete data

Solution

Use the `qqnorm` function to create the basic quantile-quantile plot; then use `qqline` to augment it with a diagonal line:

```
> qqnorm(x)
> qqline(x)
```

Here, `x` is a numeric vector.

Discussion

Sometimes it's important to know if your data is normally distributed. A quantile-quantile (Q-Q) plot is a good first check.

The `Cars93` dataset contains a Price column. Is it normally distributed? This code snippet creates a Q-Q plot of Price, shown in the lefthand panel of Figure 10-20:

```
> data(Cars93, package="MASS")
> qqnorm(Cars93$Price, main="Q-Q Plot: Price")
> qqline(Cars93$Price)
```

Figure 10-20. Quantile-quantile (Q-Q) plots

If the data had a perfect normal distribution, then the points would fall exactly on the diagonal line. Many points are close, especially in the middle section, but the points in the tails are pretty far off. Too many points are above the line, indicating a general skew to the left.

The leftward skew might be cured by a logarithmic transformation. We can plot log(Price), which yields the righthand panel of Figure 10-20:

```
> data(Cars93, package="MASS")
> qqnorm(log(Cars93$Price), main="Q-Q Plot: log(Price)")
> qqline(log(Cars93$Price))
```

Notice that the points in the new plot are much better behaved, staying close to the line except in the extreme left tail. It appears that log(Price) is approximately Normal.

See Also

See Recipe 10.22 for creating Q-Q plots for other distributions. See Recipe 11.15 for an application of Normal Q-Q plots to diagnosing linear regression.

10.22 Creating Other Quantile-Quantile Plots

Problem

You want to view a quantile-quantile plot for your data, but the data is not normally distributed.

Solution

For this recipe, you must have some idea of the underlying distribution, of course. The solution is built from the following steps:

- Use the ppoints function to generate a sequence of points between 0 and 1.
- Transform those points into quantiles, using the quantile function for the assumed distribution.
- Sort your sample data.
- Plot the sorted data against the computed quantiles.
- Use abline to plot the diagonal line.

This can all be done in two lines of R code. Here is an example that assumes your data, y, has a Student's t distribution with 5 degrees of freedom. Recall that the quantile function for Student's t is qt and that its second argument is the degrees of freedom:

```
> plot(qt(ppoints(y), 5), sort(y))
> abline(a=0, b=1)
```

Discussion

The solution looks complicated, but it's not. The R code will always look like this:

```
> plot(quantileFunction(ppoints(y), ...), sort(y))
> abline(a=0, b=1)
```

where *quantileFunction* is the quantile function for the assumed distribution and "..." are the parameters needed to *quantileFunction* beyond y. The abline function draws a line along the diagonal, where all the data points would ideally fall.

I'll illustrate this recipe by taking a random sample from an exponential distribution with a mean of 10 (or, equivalently, a rate of 1/10):

```
> RATE <- 1/10
> y <- rexp(N, rate=RATE)
```

The quantile function for the exponential distribution is qexp, which takes a rate argument. We can create the Q-Q plot in this way:

```
> plot(qexp(ppoints(y), rate=RATE), sort(y))
> abline(a=0, b=1)
```

The resulting plot would look too plain, so we include a title and labels to help the viewer. Here is the actual code that created Figure 10-21:

```
> plot(qexp(ppoints(y), rate=RATE), sort(y),
+      main="Q-Q Plot", xlab="Theoretical Quantiles", ylab="Sample Quantiles")
> abline(a=0, b=1)
```

It is interesting to see how far the Q-Q points can deviate from the diagonal—even when we know the exact distribution of the population.

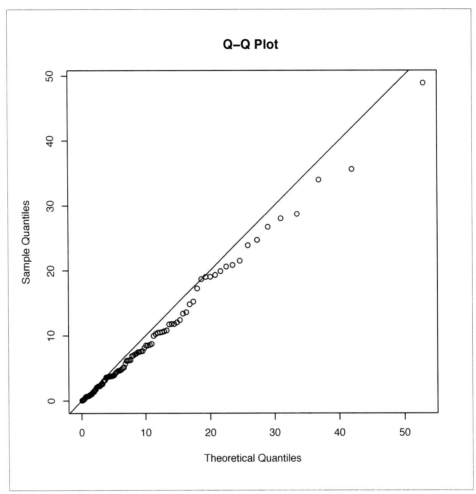

Figure 10-21. Quantile-quantile plot, exponential distribution

10.23 Plotting a Variable in Multiple Colors

Problem

You want to plot your data in multiple colors, typically to make the plot more informative, readable, or interesting.

Solution

Use the col argument of the plot function:

```
> plot(x, col=colors)
```

The value of col can be:

- One color, in which case all data points are that color.
- A vector of colors, the same length as x, in which case each value of x is colored with its corresponding color.
- A short vector, in which case the vector of colors is recycled.

Discussion

If you plot x like this, the plot is drawn in black:

```
> plot(x)
```

You can draw the entire plot in blue, like this:

```
> plot(x, col="blue")
```

However, it is much more useful (and interesting) to vary the color in a way that illuminates the data. I'll illustrate this by plotting a graphic two ways, once in black and white and once with simple shading. This call to plot produces the basic black-and-white graphic shown in the lefthand panel of Figure 10-22:

```
> plot(x, type="h", lwd=3)
```

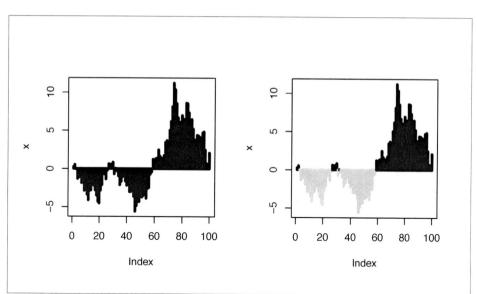

Figure 10-22. Adding shading to a plot

(The lwd=3 argument widens the lines, making them easier to see.)

The righthand panel of Figure 10-22 is produced by creating a vector of "gray" and "black" according to the sign of x and then plotting x using those colors:

```
> colors <- ifelse(x >= 0, "black", "gray")
> plot(x, type='h', lwd=3, col=colors)
```

The negative values are now plotted in gray because the corresponding element of colors is "gray".

I used grayscale shading because some editions of this book are in black and white, so colors would not be visible to the reader. You could easily use full colors instead—for example, by replacing the colors definition with this assignment statement:

```
> colors <- ifelse(x >= 0, "green", "red")
```

That would plot the positive values in green and the negative values in red.

Oddly, this recipe does not work when plotting lines. This call to plot might seem to plot a line whose colors alternate between blue and green:

```
> plot(x, type="l", col=c("blue", "green"))
```

But it doesn't work. The first color, "blue", is used for all line segments, and the second color, "green", is ignored.

See Also

See Recipe 5.3 regarding the Recycling Rule. Execute colors() to see a list of available colors, and use the segments function to plot line segments in multiple colors.

10.24 Graphing a Function

Problem

You want to graph the value of a function.

Solution

The curve function can graph a function, given the function and the limits of its domain:

```
> curve(sin, -3, +3)        # Graph the sine function from -3 to +3
```

Discussion

The lefthand panel of Figure 10-23 shows a graph of the standard normal density function. The graph was created like this:

```
> curve(dnorm, -3.5, +3.5,
+       main="Std. Normal Density")
```

The curve function calls the dnorm function for a range of arguments from –3.5 to +3.5 and then plots the result. Like any well-behaved, high-level graphics function, it accepts a main argument that specifies a title.

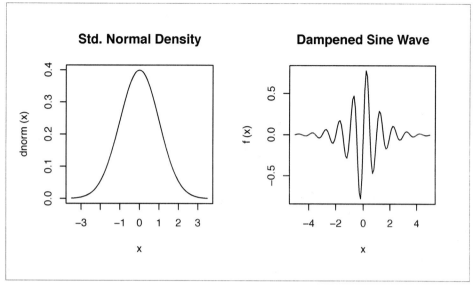

Figure 10-23. Graphing a function

curve can graph any function that takes one argument and returns one value. The righthand panel of Figure 10-23 was created by defining a local function and plotting it:

```
> f <- function(x) exp(-abs(x)) * sin(2*pi*x)
> curve(f, -5, +5, main="Dampened Sine Wave")
```

See Also

See Recipe 2.12 for how to define a function.

10.25 Pausing Between Plots

Problem

You are creating several plots, and each plot is overwriting the previous one. You want R to pause between plots so you can view each one before it's overwritten.

Solution

There is a global graphics option called ask. Set it to TRUE, and R will pause before each new plot:

```
> par(ask=TRUE)
```

When you are tired of R pausing between plots, set it to FALSE:

```
> par(ask=FALSE)
```

Discussion

When `ask` is `TRUE`, R will print this message immediately before starting a new plot:

```
Waiting to confirm page change...
```

When you are ready, click once on the graphics window. R will begin the next plot.

See Also

If one graph is overwriting another, consider using Recipe 10.26 to plot multiple graphs in one frame. See Recipe 10.29 for more about changing graphical parameters.

10.26 Displaying Several Figures on One Page

Problem

You want to display several plots side by side on one page.

Solution

First, divide the graphics window into a matrix of *N* rows and *M* columns. This is done by setting the graphics parameter called `mfrow`. Its value is a two-element vector giving the number of rows and columns:

```
> par(mfrow=(c(N,M))        # Divide the graphics window into N x M matrix
```

Second, fill each cell in the matrix by repeatedly calling `plot`. After creating a 3 × 2 grid, for example, you would call `plot` six times to fill the entire grid.

Discussion

Let's use a multifigure plot to display four different beta distributions. First, we create four plotting areas in a 2 × 2 grid:

```
> par(mfrow=c(2,2))
```

Second, we call `plot` four times, plotting a different beta distribution each time:[*]

```
> Quantile <- seq(from=0, to=1, length.out=30)
> plot(Quantile, dbeta(Quantile, 2, 4), type="l", main="First")
> plot(Quantile, dbeta(Quantile, 4, 2), type="l", main="Second")
> plot(Quantile, dbeta(Quantile, 1, 1), type="l", main="Third")
> plot(Quantile, dbeta(Quantile, 0.5, 0.5), type="l", main="Fourth")
```

The result is shown in Figure 10-24. Observe that each plot has its own title (e.g., `main="First"`), which I used here to illustrate the plotting order. Each plot could also have its own labels, legend, grid, and other adornments.

[*] Recall that the **dbeta** function computes the probability density function of the beta distribution. The second and third arguments are the distribution's shape parameters.

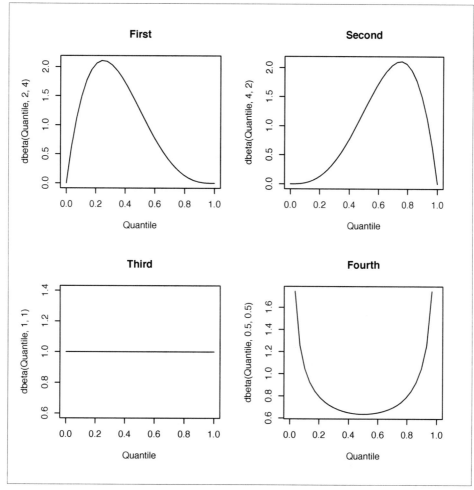

Figure 10-24. Plots of several beta distributions

When you create the graphical matrix using mfrow, R fills the graphics window row by row. In the example just given, R filled the first row and then the second row. You might prefer to fill it column by column instead. A graphics parameter called mfcol lets you do that. It works like mfrow but forces R to fill the matrix column by column. This call to par divides the graphics window into a matrix of two rows and three columns:

```
> par(mfcol=c(3,3))        # mfcol, not mfrow
```

Successive calls to plot will fill the first column, then the second column, and finally the third column.

The mfrow and mfcol parameters let you create a simple matrix of plots. For more elaborate designs, consider the layout function and the split.screen function.

See Also

Recipe 8.11 shows a similar example of plotting multiple figures. Another way to display multiple plots is by opening multiple graphics windows; see Recipe 10.27. See Recipe 10.29 for more about the par function and graphical parameters.

The grid package and the lattice package contain additional tools for multifigure layout.

10.27 Opening Additional Graphics Windows

Problem

You want to open another graphics window so you can simultaneously view two graphs.

Solution

Use the win.graph function to open a second graphics window:

```
> win.graph()
```

Subsequent graphics output will appear in the new window. Your original graphics window will remain unchanged.

Discussion

Every time you create a new graph, it overwrites the old contents of the graphics window. Sometimes you want to preserve the existing graph while creating a new graph, usually so you can compare them side by side.

The win.graph function creates a new graphics window. The new window becomes the *active window*, which means that all graphics output is directed there. All other graphics windows become *inactive*—their contents are frozen.

Eventually, you may want to freeze the active window, switch to an inactive window, and overwrite *that* window. The dev.set function switches between windows. The active window becomes inactive, and the next window (in numeric order) becomes active:

```
> dev.set()
windows
      2
```

dev.set returns the number of the now-active window.

By default, R creates a square window. You can supply width and height arguments to `win.graph`, thereby altering the default dimensions. I prefer graph dimensions based on the Golden Ratio (1.618...), for example, so I often create new graphics windows like this:[†]

```
> win.graph(width=7.0, height=5.5)
```

Finally, note that `win.graph` is a system-independent convenience function: it works on all platforms but is just a wrapper for the system-dependent function that actually creates the new graphics windows. You'll have more control if you use the system-dependent function, but your code won't be portable. See the help page for `win.graph`.

See Also

See Recipe 10.26 for creating multiple plots in one window.

10.28 Writing Your Plot to a File

Problem

You want to save your graphics in a file, such as a PNG, JPEG, or PostScript file.

Solution

Create the graphics on your screen; then call the `savePlot` function:

```
> savePlot(filename="filename.ext", type="type")
```

The *type* argument is platform specific. See the `Devices` help page for a list of types available on your machine:

```
> help(Devices)
```

Most platforms allows a *type* of "png" or "jpeg", but each platform has others, too.

Discussion

On Windows, a shortcut is to select the graphics window and then click on File → Save as... from the menu. That will prompt you for a file type and a file name before writing the file.

The procedure is similar on OS X.

Remember that the file will be written to your current working directory (unless you use an absolute file path), so be certain you know which directory that is before calling `savePlot` (Recipe 3.1).

[†] Yes, I know that 7.0/5.5 is not the Golden Ratio. The `width` and `height` arguments are the dimensions of the outer frame; owing to adjustments for margins, the (inner) plot area will be smaller. This width and height yield a plot area with the appropriate dimensions on my computer. Your mileage may vary.

The Solution will differ in a batch script:

- Call a function to open a new graphics file, such as png(...) or jpeg(...).
- Call plot and its friends to generate the graphics image.
- Call dev.off() to close the graphics file.

Functions for opening graphics files depend upon your platform and file format. Again, see help(Devices) for a list of available functions on your platform. A common function is png, which creates a PNG file; it is used like this:

```
> png("filename.ext", width=w, height=h)
```

Here *w* is the desired width and *h* is the desired height, both in pixels. Using that function, the recipe works like this to create *myPlot.png*:

```
> png("myPlot.png", width=648, height=432)     # Or whatever dimensions work for you
> plot(x, y, main="Scatterplot of X, Y")
> dev.off()
```

See Also

See Recipe 3.1 for more about the current working directory.

10.29 Changing Graphical Parameters

Problem

You want to change a global parameter of the graphics software, such as line type, background color, or font size.

Solution

Use the par function, which lets you set values of global graphics parameters. For example, this call to par will change the default line width from 1 to 2:

```
> par(lwd=2)
```

Discussion

Every graphics plot depends on many assumptions. How wide should the lines be? What's the background color? What's the foreground color? How large should the font be? How wide must the margins be? The assumed answers are given by global, graphical parameters and their default values. Thank goodness for those defaults; otherwise, the simplest plot would be a nightmare of details.

Sometimes, however, the defaults don't work for you, and you need to change them. The par function does that. There are dozens of such parameters, and serious graphics geeks will want to study the full list that is given on the help page for par.

To see a parameter's current value, call par with the parameter name given as a string:

```
> par("lty")          # What is the default line type?
[1] "solid"
> par("bg")           # What is the default background color?
[1] "transparent"
```

To set a parameter's value, pass it to par as a function parameter. Table 10-1 lists some common uses.

Table 10-1. Parameters for the par function

Parameter and type	Purpose	Example
ask=*logical*	Pause before every new graph if TRUE (Recipe 10.25)	par(ask=TRUE)
bg="*color*"	Background color	par(bg="lightyellow")
cex=*number*	Height of text and plotted points, expressed as a multiple of normal size	par(cex=1.5)
col="*color*"	Default plotting color	par(col="blue")
fg="*color*"	Foreground color	par(fg="gray")
lty="*linetype*"	Type of line: solid, dotted, dashed, etc. (Recipe 10.13)	par(lty="dotted")
lwd=*number*	Line width: 1 = normal, 2 = thicker, 3 = even thicker, etc.	par(lwd=2)
mfcol=c(*nr,nc*) or mfrow=c(*nr,nc*)	Create a multifigure plot matrix with *nr* rows and *nc* columns (Recipe 10.26)	par(mfrow=c(2,2))
new=*logical*	Used to plot one figure on top of another	par(new=TRUE)
pch=*pointtype*	Default point type (see help page for points function)	par(pch=21)
xlog=*logical*	Use logarithmic X scale	par(xlog=TRUE)
ylog=*logical*	Use logarithmic Y scale	par(ylog=TRUE)

The par function comes with a warning, however.

Changing a global parameter has a global effect, naturally. The change will affect all your plots, not just the next one, until you reverse the change. Furthermore, if you call packages that create their own graphics, the global change will also affect them. So think twice before you diddle with those globals: do you really want *all* lines in *all* graphics to be (say) magenta, dotted, and three times wider? Probably not, so use local parameters rather than global parameters whenever possible.

See Also

The help page for par lists the global graphics parameters; the chapter of *R in a Nutshell* on graphics includes the list with useful annotations. *R Graphics* contains extensive explanations of graphics parameters.

Linear Regression and ANOVA

Introduction

In statistics, modeling is where we get down to business. Models quantify the relationships between our variables. Models let us make predictions.

A *simple linear regression* is the most basic model. It's just two variables and is modeled as a linear relationship with an error term:

$$y_i = \beta_0 + \beta_1 x_i + \varepsilon_i$$

We are given the data for x and y. Our mission is to *fit the model*, which will give us the best estimates for β_0 and β_1 (Recipe 11.1).

That generalizes naturally to multiple linear regression, where we have multiple variables on the righthand side of the relationship (Recipe 11.2):

$$y_i = \beta_0 + \beta_1 u_i + \beta_2 v_i + \beta_3 w_i + \varepsilon_i$$

Statisticians call u, v, and w the *predictors* and y the *response*. Obviously, the model is useful only if there is a fairly linear relationship between the predictors and the response, but that requirement is much less restrictive than you might think. Recipe 11.11 discusses transforming your variables into a (more) linear relationship so that you can use the well-developed machinery of linear regression.

The beauty of R is that anyone can build these linear models. The models are built by a function, lm, which returns a model object. From the model object, we get the coefficients (β_i) and regression statistics. It's easy.

The horror of R is that anyone can build these models. Nothing requires you to check that the model is reasonable, much less statistically significant. Before you blindly believe a model, check it. Most of the information you need is in the regression summary (Recipe 11.4):

Is the model statistically significant?
 Check the F statistic at the bottom of the summary.

Are the coefficients significant?
> Check the coefficient's *t* statistics and *p*-values in the summary, or check their confidence intervals (Recipe 11.13).

Is the model useful?
> Check the R^2 near the bottom of the summary.

Does the model fit the data well?
> Plot the residuals and check the regression diagnostics (Recipes 11.14 and 11.15).

Does the data satisfy the assumptions behind linear regression?
> Check whether the diagnostics confirm that a linear model is reasonable for your data (Recipe 11.15).

ANOVA

Analysis of variance (ANOVA) is a powerful statistical technique. First-year graduate students in statistics are taught ANOVA almost immediately because of its importance, both theoretical and practical. I am often amazed, however, at the extent to which people outside the field are unaware of its purpose and value.

Regression creates a model, and ANOVA is one method of evaluating such models. The mathematics of ANOVA are intertwined with the mathematics of regression, so statisticians usually present them together; I follow that tradition here.

ANOVA is actually a family of techniques that are connected by a common mathematical analysis. This chapter mentions several applications:

One-way ANOVA
> This is the simplest application of ANOVA. Suppose you have data samples from several populations and are wondering whether the populations have different means. One-way ANOVA answers that question. If the populations have normal distributions, use the `oneway.test` function (Recipe 11.20); otherwise, use the nonparametric version, the `kruskal.test` function (Recipe 11.23).

Model comparison
> When you add or delete a predictor variable from a linear regression, you want to know whether that change did or did not improve the model. The `anova` function compares two regression models and reports whether they are significantly different (Recipe 11.24).

ANOVA table
> The `anova` function can also construct the ANOVA table of a linear regression model, which includes the *F* statistic needed to gauge the model's statistical significance (Recipe 11.3). This important table is discussed in nearly every textbook on regression.

The following references contain more about the mathematics of ANOVA.

See Also

There are many good texts on linear regression. One of my favorites is *Applied Linear Regression Models* (4th ed.) by Kutner, Nachtsheim, and Neter (McGraw-Hill/Irwin). I generally follow their terminology and conventions in this chapter.

I also like *Practical Regression and ANOVA Using R* by Julian Faraway because it illustrates regression using R and is quite readable. It is unpublished, but you can download the free PDF (*http://cran.r-project.org/doc/contrib/Faraway-PRA.pdf*) from CRAN.

11.1 Performing Simple Linear Regression

Problem

You have two vectors, x and y, that hold paired observations: (x_1, y_1), (x_2, y_2), ..., (x_n, y_n). You believe there is a linear relationship between x and y, and you want to create a regression model of the relationship.

Solution

The lm function performs a linear regression and reports the coefficients:

```
> lm(y ~ x)

Call:
lm(formula = y ~ x)

Coefficients:
(Intercept)            x
      17.72         3.25
```

Discussion

Simple linear regression involves two variables: a *predictor variable*, often called x; and a *response variable*, often called y. The regression uses the ordinary least-squares (OLS) algorithm to fit the linear model:

$$y_i = \beta_0 + \beta_1 x_i + \varepsilon_i$$

where β_0 and β_1 are the regression coefficients and the ε_i are the error terms.

The lm function can perform linear regression. The main argument is a *model formula*, such as y ~ x. The formula has the response variable on the left of the tilde character (~) and the predictor variable on the right. The function estimates the regression coefficients, β_0 and β_1, and reports them as the intercept and the coefficient of x, respectively:

```
Coefficients:
(Intercept)            x
      17.72         3.25
```

In this case, the regression equation is:

$$y_i = 17.72 + 3.25x_i + \varepsilon_i$$

It is quite common for data to be captured inside a data frame, in which case you want to perform a regression between two data frame columns. Here, x and y are columns of a data frame dfrm:

```
> dfrm
           x            y
1    0.04781401    5.406651
2    1.90857986   19.941568
3    2.79987246   23.922613
4    4.46755305   32.432904
5    3.76490363   44.259268
6    5.92364632   61.151480
7    8.04611587   26.305505
8    7.11097986   43.606087
9    9.73645966   58.262112
10   9.19324543   57.631029
.
. (etc.)
.
```

The lm function lets you specify a data frame by using the data parameter. If you do, the function will take the variables from the data frame and not from your workspace:

```
> lm(y ~ x, data=dfrm)        # Take x and y from dfrm

Call:
lm(formula = y ~ x, data = dfrm)

Coefficients:
(Intercept)            x
      17.72         3.25
```

11.2 Performing Multiple Linear Regression

Problem

You have several predictor variables (e.g., u, v, and w) and a response variable y. You believe there is a linear relationship between the predictors and the response, and you want to perform a linear regression on the data.

Solution

Use the lm function. Specify the multiple predictors on the righthand side of the formula, separated by plus signs (+):

```
> lm(y ~ u + v + w)
```

Discussion

Multiple linear regression is the obvious generalization of simple linear regression. It allows multiple predictor variables instead of one predictor variable and still uses OLS to compute the coefficients of a linear equation. The three-variable regression just given corresponds to this linear model:

$$y_i = \beta_0 + \beta_1 u_i + \beta_2 v_i + \beta_3 w_i + \varepsilon_i$$

R uses the lm function for both simple and multiple linear regression. You simply add more variables to the righthand side of the model formula. The output then shows the coefficients of the fitted model:

```
> lm(y ~ u + v + w)

Call:
lm(formula = y ~ u + v + w)

Coefficients:
(Intercept)           u           v           w
     1.4222      1.0359      0.9217      0.7261
```

The data parameter of lm is especially valuable when the number of variables increases, since it's much easier to keep your data in one data frame than in many separate variables. Suppose your data is captured in a data frame, such as the dfrm variable shown here:

```
> dfrm
          y            u           v           w
1   6.584519   0.79939065   2.7971413   4.366557
2   6.425215  -2.31338537   2.7836201   4.515084
3   7.830578   1.71736899   2.7570401   3.865557
4   2.757777   1.27652888   0.4191765   2.547935
5   5.794566   0.39643488   2.3785468   3.265971
6   7.314611   1.82247760   1.8291302   4.518522
7   2.533638  -1.34186107   2.3472593   2.570884
8   8.696910   0.75946803   3.4028180   4.442560
9   6.304464   0.92000133   2.0654513   2.835248
10  8.095094   1.02341093   2.6729252   3.868573
.
. (etc.)
.
```

When we supply dfrm to the data parameter of lm, R looks for the regression variables in the columns of the data frame:

```
> lm(y ~ u + v + w, data=dfrm)

Call:
lm(formula = y ~ u + v + w, data = dfrm)

Coefficients:
```

```
(Intercept)           u           v           w
       1.4222      1.0359      0.9217      0.7261
```

See Also

See Recipe 11.1 for simple linear regression.

11.3 Getting Regression Statistics

Problem

You want the critical statistics and information regarding your regression, such as R^2, the F statistic, confidence intervals for the coefficients, residuals, the ANOVA table, and so forth.

Solution

Save the regression model in a variable, say m:

```
> m <- lm(y ~ u + v + w)
```

Then use functions to extract regression statistics and information from the model:

anova(m)
 ANOVA table

coefficients(m)
 Model coefficients

coef(m)
 Same as coefficients(m)

confint(m)
 Confidence intervals for the regression coefficients

deviance(m)
 Residual sum of squares

effects(m)
 Vector of orthogonal effects

fitted(m)
 Vector of fitted y values

residuals(m)
 Model residuals

resid(m)
 Same as residuals(m)

summary(m)
 Key statistics, such as R^2, the F statistic, and the residual standard error (σ)

```
vcov(m)
```
Variance–covariance matrix of the main parameters

Discussion

When I started using R, the documentation said use the lm function to perform linear regression. So I did something like this, getting the output shown in Recipe 11.2:

```
> lm(y ~ u + v + w)

Call:
lm(formula = y ~ u + v + w)

Coefficients:
(Intercept)            u            v            w
     1.4222       1.0359       0.9217       0.7261
```

I was so disappointed! The output was nothing compared to other statistics packages such as SAS. Where is R^2? Where are the confidence intervals for the coefficients? Where is the F statistic, its p-value, and the ANOVA table?

Of course, all that information is available—you just have to ask for it. Other statistics systems dump everything and let you wade through it. R is more minimalist. It prints a bare-bones output and lets you request what more you want.

The lm function returns a *model object* that you can assign to a variable:

```
> m <- lm(y ~ u + v + w)
```

From the model object, you can extract important information using specialized functions. The most important function is summary:

```
> summary(m)

Call:
lm(formula = y ~ u + v + w)

Residuals:
    Min      1Q  Median      3Q     Max
-3.3965 -0.9472 -0.4708  1.3730  3.1283

Coefficients:
            Estimate Std. Error t value Pr(>|t|)
(Intercept)   1.4222     1.4036   1.013  0.32029
u             1.0359     0.2811   3.685  0.00106 **
v             0.9217     0.3787   2.434  0.02211 *
w             0.7261     0.3652   1.988  0.05744 .
---
Signif. codes:  0 '***' 0.001 '**' 0.01 '*' 0.05 '.' 0.1 ' ' 1

Residual standard error: 1.625 on 26 degrees of freedom
Multiple R-squared: 0.4981,     Adjusted R-squared: 0.4402
F-statistic: 8.603 on 3 and 26 DF,  p-value: 0.0003915
```

The summary shows the estimated coefficients. It shows the critical statistics, such as R^2 and the F statistic. It shows an estimate of σ, the standard error of the residuals. The summary is so important that there is an entire recipe devoted to understanding it (Recipe 11.4).

There are specialized extractor functions for other important information:

Model coefficients (point estimates)

```
> coef(m)
(Intercept)           u           v           w
  1.4222050   1.0358725   0.9217432   0.7260653
```

Confidence intervals for model coefficients

```
> confint(m)
                   2.5 %     97.5 %
(Intercept)  -1.46302727   4.307437
u             0.45805053   1.613694
v             0.14332834   1.700158
w            -0.02466125   1.476792
```

Model residuals

```
> resid(m)
          1           2           3           4           5           6
-1.41440465  1.55535335 -0.71853222 -2.22308948 -0.60201283 -0.96217874
          7           8           9          10          11          12
-1.52877080  0.12587924 -0.03313637  0.34017869  1.28200521 -0.90242817
         13          14          15          16          17          18
 2.04481731  1.13630451 -1.19766679 -0.60210494  1.79964497  1.25941264
         19          20          21          22          23          24
-2.03323530  1.40337142 -1.25605632 -0.84860707 -0.47307439 -0.76335244
         25          26          27          28          29          30
 2.16275214  1.53483492  1.65085364 -3.39647629 -0.46853750  3.12825629
```

Residual sum of squares

```
> deviance(m)
[1] 68.69616
```

ANOVA table

```
> anova(m)
Analysis of Variance Table

Response: y
          Df Sum Sq Mean Sq F value   Pr(>F)
u          1 27.916 27.9165 10.5658 0.003178 **
v          1 29.830 29.8299 11.2900 0.002416 **
w          1 10.442 10.4423  3.9522 0.057436 .
Residuals 26 68.696  2.6422
---
Signif. codes:  0 '***' 0.001 '**' 0.01 '*' 0.05 '.' 0.1 ' ' 1
```

If you find it annoying to save the model in a variable, you are welcome to use one-liners such as this:

```
> summary(lm(y ~ u + v + w))
```

See Also

See Recipe 11.4. See Recipe 11.16 for regression statistics specific to model diagnostics.

11.4 Understanding the Regression Summary

Problem

You created a linear regression model, m. However, you are confused by the output from summary(m).

Discussion

The model summary is important because it links you to the most critical regression statistics. Here is the model summary from Recipe 11.3:

```
> summary(m)

Call:
lm(formula = y ~ u + v + w)

Residuals:
    Min      1Q  Median      3Q     Max
-3.3965 -0.9472 -0.4708  1.3730  3.1283

Coefficients:
            Estimate Std. Error t value Pr(>|t|)
(Intercept)   1.4222     1.4036   1.013  0.32029
u             1.0359     0.2811   3.685  0.00106 **
v             0.9217     0.3787   2.434  0.02211 *
w             0.7261     0.3652   1.988  0.05744 .
---
Signif. codes:  0 '***' 0.001 '**' 0.01 '*' 0.05 '.' 0.1 ' ' 1

Residual standard error: 1.625 on 26 degrees of freedom
Multiple R-squared: 0.4981,     Adjusted R-squared: 0.4402
F-statistic: 8.603 on 3 and 26 DF,  p-value: 0.0003915
```

Let's dissect this summary by section. We'll read it from top to bottom—even though the most important statistic, the F statistic, appears at the end:

Call

```
Call:
lm(formula = y ~ u + v + w)
```

This shows how lm was called when it created the model, which is important for putting this summary into the proper context.

Residuals statistics

```
Residuals:
    Min      1Q  Median      3Q     Max
-3.3965 -0.9472 -0.4708  1.3730  3.1283
```

Ideally, the regression residuals would have a perfect, normal distribution. These statistics help you identify possible deviations from normality. The OLS algorithm is mathematically guaranteed to produce residuals with a mean of zero.[*] Hence the sign of the median indicates the skew's direction, and the magnitude of the median indicates the extent. In this case the median is negative, which suggests some skew to the left.

If the residuals have a nice, bell-shaped distribution, then the first quartile (1Q) and third quartile (3Q) should have about the same magnitude. In this example, the larger magnitude of 3Q versus 1Q (1.3730 versus 0.9472) indicates a slight skew to the right in our data, although the negative median makes the situation less clear-cut.

The Min and Max residuals offer a quick way to detect extreme outliers in the data, since extreme outliers (in the response variable) produce large residuals.

Coefficients

```
Coefficients:
            Estimate Std. Error t value Pr(>|t|)
(Intercept)   1.4222     1.4036   1.013  0.32029
u             1.0359     0.2811   3.685  0.00106 **
v             0.9217     0.3787   2.434  0.02211 *
w             0.7261     0.3652   1.988  0.05744 .
---
Signif. codes:  0 '***' 0.001 '**' 0.01 '*' 0.05 '.' 0.1 ' ' 1
```

The column labeled `Estimate` contains the estimated regression coefficients as calculated by ordinary least squares.

Theoretically, if a variable's coefficient is zero then the variable is worthless; it adds nothing to the model. Yet the coefficients shown here are only estimates, and they will never be exactly zero. We therefore ask: Statistically speaking, how likely is it that the true coefficient is zero? That is the purpose of the *t* statistics and the *p*-values, which in the summary are labeled (respectively) `t value` and `Pr(>|t|)`.

The *p*-value is a probability. It gauges the likelihood that the coefficient is *not* significant, so smaller is better. Big is bad because it indicates a high likelihood of insignificance. In this example, the *p*-value for the *u* coefficient is a mere 0.00106, so *u* is likely significant. The *p*-value for *w*, however, is 0.05744; this is just over

[*] Unless you performed the linear regression without an intercept term (Recipe 11.5).

our conventional limit of 0.05, which suggests that w is likely insignificant.[†] Variables with large p-values are candidates for elimination.

A handy feature is that R flags the significant variables for quick identification. Do you notice the extreme righthand column containing double asterisks (**), a single asterisk (*), and a period(.)? That column highlights the significant variables. The line labeled "Signif. codes" at the bottom gives a cryptic guide to the flags' meanings:

***	p-value between 0 and 0.001
**	p-value between 0.001 and 0.01
*	p-value between 0.01 and 0.05
.	p-value between 0.05 and 0.1
(blank)	p-value between 0.1 and 1.0

The column labeled Std. Error is the standard error of the estimated coefficient. The column labeled t value is the t statistic from which the p-value was calculated.

Residual standard error

```
Residual standard error: 1.625 on 26 degrees of freedom
```

This reports the standard error of the residuals (σ)—that is, the sample standard deviation of ε.

R^2 (coefficient of determination)

```
Multiple R-squared: 0.4981,     Adjusted R-squared: 0.4402
```

R^2 is a measure of the model's quality. Bigger is better. Mathematically, it is the fraction of the variance of y that is explained by the regression model. The remaining variance is not explained by the model, so it must be due to other factors (i.e., unknown variables or sampling variability). In this case, the model explains 0.4981 (49.81%) of the variance of y, and the remaining 0.5019 (50.19%) is unexplained.

That being said, I strongly suggest using the adjusted rather than the basic R^2. The adjusted value accounts for the number of variables in your model and so is a more realistic assessment of its effectiveness. In this case, then, I would use 0.4402, not 0.4981.

F statistic

```
F-statistic: 8.603 on 3 and 26 DF,  p-value: 0.0003915
```

The F statistic tells you whether the model is significant or insignificant. The model is significant if any of the coefficients are nonzero (i.e., if $\beta_i \neq 0$ for some i). It is insignificant if all coefficients are zero ($\beta_1 = \beta_2 = \ldots = \beta_n = 0$).

[†] The significance level of $\alpha = 0.05$ is the convention observed in this book. Your application might instead use $\alpha = 0.10$, $\alpha = 0.01$, or some other value. See the "Introduction" to Chapter 9.

Conventionally, a p-value of less than 0.05 indicates that the model is likely significant (one or more β_i are nonzero) whereas values exceeding 0.05 indicate that the model is likely not significant. Here, the probability is only 0.000391 that our model is insignificant. That's good.

Most people look at the R^2 statistic first. The statistician wisely starts with the F statistic, for if the model is not significant then nothing else matters.

See Also

See Recipe 11.3 for more on extracting statistics and information from the model object.

11.5 Performing Linear Regression Without an Intercept

Problem

You want to perform a linear regression, but you want to force the intercept to be zero.

Solution

Add "+ 0" to the righthand side of your regression formula. That will force lm to fit the model with a zero intercept:

```
> lm(y ~ x + 0)
```

The corresponding regression equation is:

$$y_i = \beta x_i + \varepsilon_i$$

Discussion

Linear regression ordinarily includes an intercept term, so that is the default in R. In rare cases, however, you may want to fit the data while assuming that the intercept is zero. In this you make a modeling assumption: when x is zero, y should be zero.

When you force a zero intercept, the lm output includes a coefficient for x but no intercept for y, as shown here:

```
> lm(y ~ x + 0)

Call:
lm(formula = y ~ x + 0)

Coefficients:
    x
2.582
```

I strongly suggest you check that modeling assumption before proceeding. Perform a regression with an intercept; then see if the intercept could plausibly be zero. Check the intercept's confidence interval. In this example, the confidence interval is (–12.17, 10.17):

```
> confint(lm(y ~ x))
                2.5 %    97.5 %
(Intercept) -12.171368 10.170070
x             2.012768  3.248262
```

Because the confidence interval contains zero, it is statistically plausible that the intercept could be zero. Now it is reasonable to rerun the regression while forcing a zero intercept.

11.6 Performing Linear Regression with Interaction Terms

Problem

You want to include an interaction term in your regression.

Solution

The R syntax for regression formulas lets you specify interaction terms. The interaction of two variables, u and v, is indicated by separating their names with an asterisk (*):

```
> lm(y ~ u*v)
```

This corresponds to the model $y_i = \beta_0 + \beta_1 u_i + \beta_2 v_i + \beta_3 u_i v_i + \varepsilon_i$, which includes the first-order interaction term $\beta_3 u_i v_i$.

Discussion

In regression, an *interaction* occurs when the product of two predictor variables is also a significant predictor (i.e., in addition to the predictor variables themselves). Suppose we have two predictors, u and v, and want to include their interaction in the regression. This is expressed by the following equation:

$$y_i = \beta_0 + \beta_1 u_i + \beta_2 v_i + \beta_3 u_i v_i + \varepsilon_i$$

Here the product term, $\beta_3 u_i v_i$, is called the *interaction term*. The R formula for that equation is:

```
y ~ u*v
```

When you write y ~ u*v, R automatically includes u, v, and their product in the model. This is for a good reason. If a model includes an interaction term, such as $\beta_3 u_i v_i$, then regression theory tells us the model should also contain the constituent variables u_i and v_i.

Likewise, if you have three predictors (u, v, and w) and want to include all their interactions, separate them by asterisks:

```
y ~ u*v*w
```

This corresponds to the regression equation:

$$y_i = \beta_0 + \beta_1 u_i + \beta_2 v_i + \beta_3 w_i + \beta_4 u_i v_i + \beta_5 u_i w_i + \beta_6 v_i w_i + \beta_7 u_i v_i w_i + \varepsilon_i$$

Now we have all the first-order interactions and a second-order interaction ($\beta_7 u_i v_i w_i$).

Sometimes, however, you may not want every possible interaction. You can explicitly specify a single product by using the colon operator (:). For example, u:v:w denotes the product term $\beta u_i v_i w_i$ but without all possible interactions. So the R formula:

```
y ~ u + v + w + u:v:w
```

corresponds to the regression equation:

$$y_i = \beta_0 + \beta_1 u_i + \beta_2 v_i + \beta_3 w_i + \beta_4 u_i v_i w_i + \varepsilon_i$$

It might seem odd that colon (:) means pure multiplication while asterisk (*) means both multiplication and inclusion of constituent terms. Again, this is because we normally incorporate the constituents when we include their interaction, so making that the default for asterisk makes sense.

There is some additional syntax for easily specifying many interactions:

(u + v + ... + w)^2
 Include all variables (u, v, ..., w) and all their first-order interactions.

(u + v + ... + w)^3
 Include all variables, all their first-order interactions, and all their second-order interactions.

(u + v + ... + w)^4
 And so forth.

Both the asterisk (*) and the colon (:) follow a "distributive law", so the following notations are also allowed:

x*(u + v + ... + w)
 Same as x*u + x*v + ... + x*w (which is the same as x + u + v + ... + w + x:u + x:v + ... + x:w).

x:(u + v + ... + w)
 Same as x:u + x:v + ... + x:w.

All this syntax gives you some flexibility in writing your formula. For example, these three formulas are equivalent:

```
y ~ u*v
y ~ u + v + u:v
y ~ (u + v)^2
```

They all define the same regression equation, $y_i = \beta_0 + \beta_1 u_i + \beta_2 v_i + \beta_3 u_i v_i + \varepsilon_i$.

See Also

The full syntax for formulas is richer than described here. See *R in a Nutshell* (O'Reilly) or the *R Language Definition* for more details.

11.7 Selecting the Best Regression Variables

Problem

You are creating a new regression model or improving an existing model. You have the luxury of many regression variables, and you want to select the best subset of those variables.

Solution

The step function can perform stepwise regression, either forward or backward. *Backward stepwise regression* starts with many variables and removes the underperformers:

```
> full.model <- lm(y ~ x1 + x2 + x3 + x4)
> reduced.model <- step(full.model, direction="backward")
```

Forward stepwise regression starts with a few variables and adds new ones to improve the model until it cannot be improved further:

```
> min.model <- lm(y ~ 1)
> fwd.model <- step(min.model, direction="forward", scope=( ~ x1 + x2 + x3 + x4 ))
```

Discussion

When you have many predictors, it can be quite difficult to choose the best subset. Adding and removing individual variables affects the overall mix, so the search for "the best" can become tedious.

The step function automates that search. Backward stepwise regression is the easiest approach. Start with a model that includes all the predictors. We call that the *full model*. The model summary, shown here, indicates that not all predictors are statistically significant:

```
> full.model <- lm(y ~ x1 + x2 + x3 + x4)
> summary(full.model)

Call:
lm(formula = y ~ x1 + x2 + x3 + x4)

Residuals:
    Min      1Q  Median      3Q     Max
-34.405  -5.153   2.025   6.525  19.186

Coefficients:
            Estimate Std. Error t value Pr(>|t|)
(Intercept)   10.274      4.573   2.247   0.0337 *
x1          -102.298     47.558  -2.151   0.0413 *
x2             4.362     40.237   0.108   0.9145
x3            75.115     34.236   2.194   0.0377 *
x4            26.286     42.239   0.622   0.5394
---
Signif. codes:  0 '***' 0.001 '**' 0.01 '*' 0.05 '.' 0.1 ' ' 1
```

```
Residual standard error: 12.19 on 25 degrees of freedom
Multiple R-squared: 0.8848,    Adjusted R-squared: 0.8664
F-statistic:    48 on 4 and 25 DF,  p-value: 2.235e-11
```

We want to eliminate the insignificant variables, so we use step to incrementally elim-inate the underperformers. The result is called the *reduced model*:

```
> reduced.model <- step(full.model, direction="backward")
Start:  AIC=154.58
y ~ x1 + x2 + x3 + x4

        Df Sum of Sq    RSS    AIC
- x2     1      1.75 3718.5 152.60
- x4     1     57.58 3774.3 153.04
<none>               3716.7 154.58
- x1     1    687.89 4404.6 157.68
- x3     1    715.67 4432.4 157.87

Step:  AIC=152.6
y ~ x1 + x3 + x4

        Df Sum of Sq    RSS    AIC
- x4     1     77.82 3796.3 151.22
<none>               3718.5 152.60
- x3     1    737.39 4455.9 156.02
- x1     1    787.96 4506.4 156.36

Step:  AIC=151.22
y ~ x1 + x3

        Df Sum of Sq    RSS    AIC
<none>               3796.3 151.22
- x1     1    884.44 4680.7 155.50
- x3     1    964.24 4760.5 156.01
```

The output from step shows the sequence of models that it explored. In this case, step removed x2 and x4 and left only x1 and x3 in the final (reduced) model. The sum-mary of the reduced model shows that it contains only significant predictors:

```
> summary(reduced.model)

Call:
lm(formula = y ~ x1 + x3)

Residuals:
    Min      1Q  Median      3Q     Max
-36.192  -2.677   2.049   7.453  19.704

Coefficients:
            Estimate Std. Error t value Pr(>|t|)
(Intercept)   10.267      4.437   2.314   0.0285 *
x1           -79.265     31.604  -2.508   0.0185 *
x3            82.726     31.590   2.619   0.0143 *
---
Signif. codes:  0 '***' 0.001 '**' 0.01 '*' 0.05 '.' 0.1 ' ' 1
```

```
Residual standard error: 11.86 on 27 degrees of freedom
Multiple R-squared: 0.8823,     Adjusted R-squared: 0.8736
F-statistic: 101.2 on 2 and 27 DF,  p-value: 2.842e-13
```

Backward stepwise regression is easy, but sometimes it's not feasible to start with "everything" because you have too many candidate variables. In that case use forward stepwise regression, which will start with nothing and incrementally add variables that improve the regression. It stops when no further improvement is possible.

A model that "starts with nothing" may look odd at first:

```
> min.model <- lm(y ~ 1)
```

This is a model with a response variable (y) but no predictor variables. (All the fitted values for y are simply the mean of y, which is what you would guess if no predictors were available.)

We must tell step which candidate variables are available for inclusion in the model. That is the purpose of the scope argument. The scope is a formula with nothing on the lefthand side of the tilde (~) and candidate variables on the righthand side:

```
> fwd.model <- step(min.model,
+                    direction="forward",
+                    scope=( ~ x1 + x2 + x3 + x4),
+                    trace=0 )
```

Here we see that x1, x2, x3, and x4 are all candidates for inclusion. (I also included trace=0 to inhibit the voluminous output from step.) The resulting model has two significant predictors and no insignificant predictors:

```
> summary(fwd.model)

Call:
lm(formula = y ~ x3 + x1)

Residuals:
    Min      1Q  Median      3Q     Max
-36.192  -2.677   2.049   7.453  19.704

Coefficients:
            Estimate Std. Error t value Pr(>|t|)
(Intercept)   10.267      4.437   2.314   0.0285 *
x3            82.726     31.590   2.619   0.0143 *
x1           -79.265     31.604  -2.508   0.0185 *
---
Signif. codes:  0 '***' 0.001 '**' 0.01 '*' 0.05 '.' 0.1 ' ' 1

Residual standard error: 11.86 on 27 degrees of freedom
Multiple R-squared: 0.8823,     Adjusted R-squared: 0.8736
F-statistic: 101.2 on 2 and 27 DF,  p-value: 2.842e-13
```

The step-forward algorithm reached the same model as the step-backward model by including x1 and x3 but excluding x2 and x4. This is a toy example, so that is not surprising. In real applications, I suggest trying both the forward and the backward regression and then comparing the results. You might be surprised.

Finally, don't get carried away by stepwise regression. It is not a panacea, it cannot turn junk into gold, and it is definitely not a substitute for choosing predictors carefully and wisely. You might think: "Oh boy! I can generate every possible interaction term for my model, then let step choose the best ones! What a model I'll get!" You'd be thinking of something like this:

```
> full.model <- lm(y ~ (x1 + x2 + x3 + x4)^4)          # All possible interactions
> reduced.model <- step(full.model, direction="backward")
```

This does not work well: most of the interaction terms are meaningless. The step function becomes overwhelmed, and you are left with many insignificant terms.

See Also

See Recipe 11.24.

11.8 Regressing on a Subset of Your Data

Problem

You want to fit a linear model to a subset of your data, not to the entire dataset.

Solution

The lm function has a subset parameter that specifies which data elements should be used for fitting. The parameter's value can be any index expression that could index your data. This shows a fitting that uses only the first 100 observations:

```
> lm(y ~ x, subset=1:100)          # Use only x[1:100]
```

Discussion

You will often want to regress only a subset of your data. This can happen, for example, when using in-sample data to create the model and out-of-sample data to test it.

The lm function has a parameter, subset, that selects the observations used for fitting. The value of subset is a vector. It can be a vector of index values, in which case lm selects only the indicated observations from your data. It can also be a logical vector, the same length as your data, in which case lm selects the observations with a corresponding TRUE.

Suppose you have 1,000 observations of (x, y) pairs and want to fit your model using only the first half of those observations. Use a subset parameter of 1:500, indicating lm should use observations 1 through 500:

```
> lm(y ~ x, subset=1:500)
```

More generally, you can use the expression `1:floor(length(x)/2)` to select the first half of your data, regardless of size:

```
> lm(y ~ x, subset=1:floor(length(x)/2))
```

Let's say your data was collected in several labs and you have a factor, `lab`, that identifies the lab of origin. You can limit your regression to observations collected in New Jersey by using a logical vector that is TRUE only for those observations:

```
> lm(y ~ x, subset=(lab == "NJ"))
```

11.9 Using an Expression Inside a Regression Formula

Problem

You want to regress on calculated values, not simple variables, but the syntax of a regression formula seems to forbid that.

Solution

Embed the expressions for the calculated values inside the I(...) operator. That will force R to calculate the expression and use the calculated value for the regression.

Discussion

If you want to regress on the sum of u and v, then this is your regression equation:

$$y_i = \beta_0 + \beta_1(u_i + v_i) + \varepsilon_i$$

How do you write that equation as a regression formula? This won't work:

```
> lm(y ~ u + v)           # Not quite right
```

Here R will interpret u and v as two separate predictors, each with its own regression coefficient. Likewise, suppose your regression equation is:

$$y_i = \beta_0 + \beta_1 u_i + \beta_2 u_i^2 + \varepsilon_i$$

This won't work:

```
> lm(y ~ u + u^2)         # That's an interaction, not a quadratic term
```

R will interpret u^2 as an interaction term (Recipe 11.6) and not as the square of u.

The solution is to surround the expressions by the I(...) operator, which prevents the expressions from being interpreted as a regression formula. Instead, it forces R to calculate the expression's value and then incorporate that value directly into the regression. Thus the first example becomes:

```
> lm(y ~ I(u + v))
```

In response to that command, R computes $u + v$ and then regresses y on the sum.

For the second example we use:

```
> lm(y ~ u + I(u^2))
```

Here R computes the square of u and then regresses on the sum $u + u^2$.

All the basic binary operators (+, -, *, /, ^) have special meanings inside a regression formula. For this reason, you must use the I(...) operator whenever you incorporate calculated values into a regression.

A beautiful aspect of these embedded transformations is that R remembers the transformations and applies them when you make predictions from the model. Consider the quadratic model described by the second example. It uses u and u^2, but we supply the value of u only and R does the heavy lifting. We don't need to calculate the square of u ourselves:

```
> m <- lm(y ~ u + I(u^2))
> predict(m, newdata=data.frame(u=13.4))     # R plugs u and u^2 into the calculation
        1
11531.59
```

See Also

See Recipe 11.10 for the special case of regression on a polynomial. See Recipe 11.11 for incorporating other data transformations into the regression.

11.10 Regressing on a Polynomial

Problem

You want to regress y on a polynomial of x.

Solution

Use the poly(x,n) function in your regression formula to regress on an n-degree polynomial of x. This example models y as a cubic function of x:

```
> lm(y ~ poly(x,3,raw=TRUE))
```

The example's formula corresponds to the following cubic regression equation:

$$y_i = \beta_0 + \beta_1 x_i + \beta_2 x_i^2 + \beta_3 x_i^3 + \varepsilon_i$$

Discussion

When I first used a polynomial model in R, I did something clunky like this:

```
> x_sq <- x^2
> x_cub <- x^3
> m <- lm(y ~ x + x_sq + x_cub)
```

Obviously, this was quite annoying, and it littered my workspace with extra variables.

It's much easier to write:

```
> m <- lm(y ~ poly(x,3,raw=TRUE))
```

The raw=TRUE is necessary. Without it, the poly function computes orthogonal polynomials instead of simple polynomials.

Beyond the convenience, a huge advantage is that R will calculate all those powers of x when you make predictions from the model (Recipe 11.18). Without that, you are stuck calculating x^2 and x^3 yourself every time you employ the model.

Here is another good reason to use poly. You cannot write your regression formula in this way:

```
> lm(y ~ x + x^2 + x^3)      # Does not do what you think!
```

R will interpret x^2 and x^3 as interaction terms, not as powers of x. The resulting model is a one-term linear regression, completely unlike your expectation. You could write the regression formula like this:

```
> lm(y ~ x + I(x^2) + I(x^3))
```

But that's getting pretty verbose. Just use poly.

See Also

See Recipe 11.6 for more about interaction terms. See Recipe 11.11 for other transformations on regression data.

11.11 Regressing on Transformed Data

Problem

You want to build a regression model for x and y, but they do not have a linear relationship.

Solution

You can embed the needed transformation inside the regression formula. If, for example, y must be transformed into $\log(y)$, then the regression formula becomes:

```
> lm(log(y) ~ x)
```

Discussion

A critical assumption behind the lm function for regression is that the variables have a linear relationship. To the extent this assumption is false, the resulting regression becomes meaningless.

Fortunately, many datasets can be transformed into a linear relationship before applying lm. Figure 11-1 shows an example of exponential decay. The left panel shows the

original data, z. The dotted line shows a linear regression on the original data; clearly, it's a lousy fit. If the data is really exponential, then a possible model is:

$$z = \exp[\beta_0 + \beta_1 t + \varepsilon]$$

where t is time and $\exp[\cdot]$ is the exponential function (e^x). This is not linear, of course, but we can linearize it by taking logarithms:

$$\log(z) = \beta_0 + \beta_1 t + \varepsilon$$

In R, that regression is simple because we can embed the log transform directly into the regression formula:

```
> m <- lm(log(z) ~ t)
> summary(m)

Call:
lm(formula = log(z) ~ t)

Residuals:
     Min       1Q   Median       3Q      Max
-0.17366 -0.09070 -0.03761  0.08080  0.36454

Coefficients:
             Estimate Std. Error t value Pr(>|t|)
(Intercept) -0.22059    0.01653  -13.35   <2e-16 ***
t           -1.16110    0.02056  -56.46   <2e-16 ***
---
Signif. codes:  0 '***' 0.001 '**' 0.01 '*' 0.05 '.' 0.1 ' ' 1

Residual standard error: 0.1275 on 98 degrees of freedom
Multiple R-squared: 0.9702,     Adjusted R-squared: 0.9699
F-statistic:  3188 on 1 and 98 DF,  p-value: < 2.2e-16
```

The right panel of Figure 11-1 shows the plot of $\log(z)$ versus time. Superimposed on that plot is their regression line. The fit appears to be much better; this is confirmed by the $R^2 = 0.97$, compared with 0.76 for the linear regression on the original data.

You can embed other functions inside your formula. If you thought the relationship was quadratic, you could use a square-root transformation:

```
> lm(sqrt(y) ~ month)
```

You can apply transformations to variables on both sides of the formula, of course. This formula regresses y on the square root of x:

```
> lm(y ~ sqrt(x))
```

This regression is for a log-log relationship between x and y:

```
> lm(log(y) ~ log(x))
```

See Also

See Recipe 11.12.

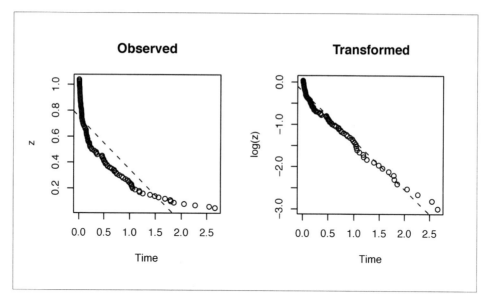

Figure 11-1. Transforming data into a linear relationship

11.12 Finding the Best Power Transformation (Box–Cox Procedure)

Problem

You want to improve your linear model by applying a power transformation to the response variable.

Solution

Use the Box–Cox procedure, which is implemented by the `boxcox` function of the MASS package. The procedure will identify a power, λ, such that transforming y into y^λ will improve the fit of your model:

```
> library(MASS)
> m <- lm(y ~ x)
> boxcox(m)
```

Discussion

To illustrate the Box–Cox transformation, let's create some artificial data using the equation $y^{-1.5} = x + \varepsilon$, where ε is an error term:

```
> x <- 10:100
> eps <- rnorm(length(x), sd=5)
> y <- (x + eps)^(-1/1.5)
```

Then we will (mistakenly) model the data using a simple linear regression and derive an adjusted R^2 of 0.6935:

```
> m <- lm(y ~ x)
> summary(m)

Call:
lm(formula = y ~ x)

Residuals:
      Min        1Q    Median        3Q       Max
-0.038866 -0.016128 -0.004948  0.008367  0.083354

Coefficients:
              Estimate Std. Error t value Pr(>|t|)
(Intercept)  1.580e-01  5.710e-03   27.67   <2e-16 ***
x           -1.340e-03  9.368e-05  -14.30   <2e-16 ***
---
Signif. codes:  0 '***' 0.001 '**' 0.01 '*' 0.05 '.' 0.1 ' ' 1

Residual standard error: 0.02347 on 89 degrees of freedom
Multiple R-squared: 0.6969,     Adjusted R-squared: 0.6935
F-statistic: 204.6 on 1 and 89 DF,  p-value: < 2.2e-16
```

When plotting the residuals against the fitted values, we get a clue that something is wrong:

```
> plot(m, which=1)     # Plot only the fitted vs residuals
```

The resulting plot, shown in the left panel of Figure 11-2, has a parabolic shape. A possible fix is a power transformation on y, so we run the Box–Cox procedure:

```
> library(MASS)
> bc <- boxcox(m)
```

The boxcox function creates a plot, shown in the right panel of Figure 11-2, that plots values of λ against the log-likelihood of the resulting model. We want to maximize that log-likelihood, so the function draws a line at the best value and also draws lines at the limits of its confidence interval. In this case, it looks like the best value is around –1.5, with a confidence interval of about (–1.75, –1.25).

Oddly, the boxcox function does not return the best value of λ. Rather, it returns the (x, y) pairs displayed in the plot. It's pretty easy to find the values of λ that yield the largest log-likelihood y. We use the which.max function:

```
> which.max(bc$y)
[1] 13
```

Then this gives us the position of the corresponding λ:

```
> lambda <- bc$x[which.max(bc$y)]
> lambda
[1] -1.515152
```

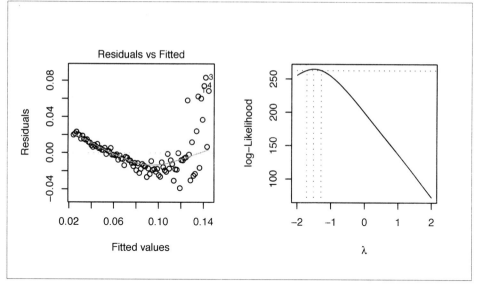

Figure 11-2. Box–Cox plots

The function reports that the best λ is –1.515. In an actual application, I would urge you to interpret this number and choose the power that makes sense to you—rather than blindly accepting this "best" value. Use the graph to assist you in that interpretation. Here, I'll go with –1.515.

We can apply the power transform to y and then fit the revised model; this gives a much better R^2 of 0.9595:

```
> z <- y^lambda
> m2 <- lm(z ~ x)
> summary(m2)

Call:
lm(formula = z ~ x)

Residuals:
    Min      1Q  Median      3Q     Max
-12.4325 -3.6608 -0.8793  2.3856 15.0977

Coefficients:
            Estimate Std. Error t value Pr(>|t|)
(Intercept)  0.61801    1.36496   0.453    0.652
x            1.03382    0.02239  46.164   <2e-16 ***
---
Signif. codes:  0 '***' 0.001 '**' 0.01 '*' 0.05 '.' 0.1 ' ' 1

Residual standard error: 5.612 on 89 degrees of freedom
Multiple R-squared: 0.9599,    Adjusted R-squared: 0.9595
F-statistic:  2131 on 1 and 89 DF,  p-value: < 2.2e-16
```

For those who prefer one-liners, the transformation can be embedded right into the revised regression formula:

```
> m2 <- lm(I(y^lambda) ~ x)
```

By default, boxcox searches for values of λ in the range –2 to +2. You can change that via the lambda argument; see the help page for details.

I suggest viewing the Box–Cox result as a starting point, not as a definitive answer. If the confidence interval for λ includes 1.0, it may be that no power transformation is actually helpful. As always, inspect the residuals before and after the transformation. Did they really improve?

See Also

See Recipes 11.11 and 11.15.

11.13 Forming Confidence Intervals for Regression Coefficients

Problem

You are performing linear regression and you need the confidence intervals for the regression coefficients.

Solution

Save the regression model in an object; then use the confint function to extract confidence intervals:

```
> m <- lm(y ~ x1 + x2)
> confint(m)
```

Discussion

The Solution uses the model $y = \beta_0 + \beta_1(x_1)_i + \beta_2(x_2)_i + \varepsilon_i$. The confint function returns the confidence intervals for the intercept (β_0), the coefficient of x_1 (β_1), and the coefficient of x_2 (β_2):

```
> confint(m)
                2.5 %    97.5 %
(Intercept) -5.332195 12.361525
x1          -3.009250  1.282295
x2           1.892837  7.093099
```

By default, confint uses a confidence level of 95%. Use the level parameter to select a different level:

```
> confint(m, level=0.99)
                0.5 %    99.5 %
```

```
(Intercept) -8.4316652 15.460995
x1          -3.7610142  2.034059
x2           0.9818897  8.004046
```

See Also

The `coefplot` function of the `arm` package can plot confidence intervals for regression coefficients.

11.14 Plotting Regression Residuals

Problem

You want a visual display of your regression residuals.

Solution

You can plot the model object by selecting the residuals plot from the available plots:

```
> m <- lm(y ~ x)
> plot(m, which=1)
```

Discussion

Normally, plotting a regression model object produces several diagnostic plots. You can select just the residuals plot by specifying `which=1`.

Figure 11-3 shows a plot of the residuals from Recipe 11.1. R draws a smoothed line through the residuals as a visual aid to finding significant patterns—for example, a slope or a parabolic shape.

See Also

See Recipe 11.15, which contains examples of residuals plots and other diagnostic plots.

11.15 Diagnosing a Linear Regression

Problem

You have performed a linear regression. Now you want to verify the model's quality by running diagnostic checks.

Solution

Start by plotting the model object, which will produce several diagnostic plots:

```
> m <- lm(y ~ x)
> plot(m)
```

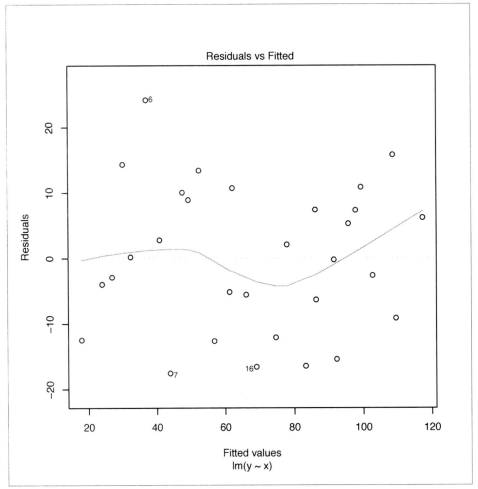

Figure 11-3. Plotting regression residuals

Next, identify possible outliers either by looking at the diagnostic plot of the residuals or by using the `outlier.test` function of the `car` package:

```
> library(car)
> outlier.test(m)
```

Finally, identify any overly influential observations (Recipe 11.16).

Discussion

R fosters the impression that linear regression is easy: just use the `lm` function. Yet fitting the data is only the beginning. It's your job to decide whether the fitted model actually works and works well.

Before anything else, you must have a statistically significant model. Check the F statistic from the model summary (Recipe 11.4) and be sure that the *p*-value is small enough for your purposes. Conventionally, it should be less than 0.05 or else your model is likely meaningless.

Simply plotting the model object produces several useful diagnostic plots:

```
> m <- lm(y ~ x)
> plot(m)
```

Figure 11-4 shows diagnostic plots for a pretty good regression:

- The points in the Residuals vs Fitted plot are randomly scattered with no particular pattern.
- The points in the Normal Q–Q plot are more-or-less on the line, indicating that the residuals follow a normal distribution.
- In both the Scale–Location plot and the Residuals vs Leverage plots, the points are in a group with none too far from the center.

In contrast, Figure 11-5 shows the diagnostics for a not-so-good regression. Observe that the Residuals vs Fitted plot has a definite parabolic shape. This tells us that the model is incomplete: a quadratic factor is missing that could explain more variation in *y*. Other patterns in residuals are suggestive of additional problems: a cone shape, for example, may indicate nonconstant variance in *y*. Interpreting those patterns is a bit of an art, so I suggest reviewing a good book on linear regression while evaluating the plot of residuals.

There are other problems with the not-so-good diagnostics. The Normal Q–Q plot has more points off the line than it does for the good regression. Both the Scale–Location and Residuals vs Leverage plots show points scattered away from the center, which suggests that some points have excessive leverage.

Another pattern is that point number 28 sticks out in every plot. This warns us that something is odd with that observation. The point could be an outlier, for example. We can check that hunch with the outlier.test function of the car package:

```
> outlier.test(m)

max|rstudent| = 3.183304, degrees of freedom = 27,
unadjusted p = 0.003648903, Bonferroni p = 0.1094671

Observation: 28
```

The outlier.test identifies the model's most outlying observation. In this case, it identified observation number 28 and so confirmed that it could be an outlier.

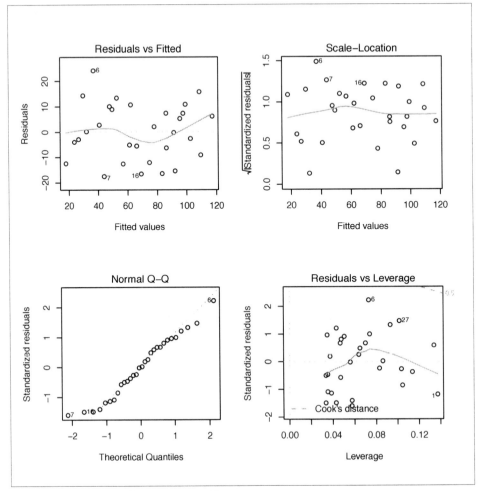

Figure 11-4. Diagnostic plots—pretty good fit

See Also

See Recipes 11.4 and 11.16. The car package is not part of the standard distribution of R; see Recipe 3.9.

11.16 Identifying Influential Observations

Problem

You want to identify the observations that are having the most influence on the regression model. This is useful for diagnosing possible problems with the data.

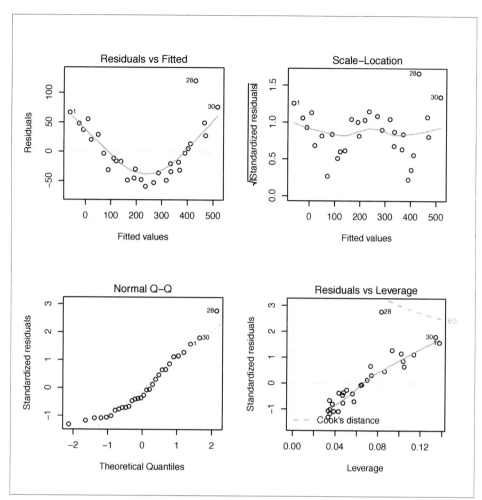

Figure 11-5. Diagnostic plots—not-so-good fit

Solution

The `influence.measures` function reports several useful statistics for identifying influential observations, and it flags the significant ones with an asterisk (*). Its main argument is the model object from your regression:

```
> influence.measures(m)
```

Discussion

The title of this recipe could be "Identifying *Overly* Influential Observations", but that would be redundant. All observations influence the regression model, even if only a little. When a statistician says that an observation is *influential*, it means that removing

the observation would significantly change the fitted regression model. We want to identify those observations because they might be outliers that distort our model; we owe it to ourselves to investigate them.

The `influence.measures` function reports several statistics: DFBETAS, DFFITS, covariance ratio, Cook's distance, and hat matrix values. If any of these measures indicate that an observation is influential, the function flags that observation with an asterisk (*) along the righthand side:

```
> influence.measures(m)
Influence measures of
        lm(formula = z ~ x) :

    dfb.1_    dfb.x   dffit cov.r   cook.d    hat inf
1   0.64663 -0.56267  0.6466  1.04 1.98e-01 0.1373
2   0.40133 -0.33798  0.4020  1.11 8.01e-02 0.1137
3   0.29023 -0.23992  0.2914  1.14 4.29e-02 0.1035
4   0.13917 -0.11036  0.1409  1.16 1.02e-02 0.0862
5   0.40997 -0.33125  0.4133  1.05 8.34e-02 0.0932
.
. (etc.)
.
25 -0.00985  0.02449  0.0338  1.16 5.93e-04 0.0700
26 -0.02698  0.06341  0.0856  1.15 3.78e-03 0.0740
27 -0.16813  0.31877  0.3888  1.09 7.47e-02 0.1017
28 -0.34708  0.74122  0.9580  0.62 3.46e-01 0.0830   *
29 -0.09644  0.18050  0.2187  1.17 2.44e-02 0.1045
30 -0.37679  0.64112  0.7398  0.97 2.51e-01 0.1339
```

This is the model from Recipe 11.15, where we suspected that observation 28 was an outlier. An asterisk is flagging that observation, confirming that it's overly influential.

This recipe can identify influential observations, but you shouldn't reflexively delete them. Some judgment is required here. Are those observations improving your model or damaging it?

See Also

See Recipe 11.15. Use `help(influence.measures)` to get a list of influence measures and some related functions. See a regression textbook for interpretations of the various influence measures.

11.17 Testing Residuals for Autocorrelation (Durbin–Watson Test)

Problem

You have performed a linear regression and want to check the residuals for autocorrelation.

Solution

The Durbin—Watson test can check the residuals for autocorrelation. The test is implemented by the dwtest function of the lmtest package:

```
> library(lmtest)
> m <- lm(y ~ x)          # Create a model object
> dwtest(m)               # Test the model residuals
```

The output includes a *p*-value. Conventionally, if $p < 0.05$ then the residuals are significantly correlated whereas $p > 0.05$ provides no evidence of correlation.

You can perform a visual check for autocorrelation by graphing the autocorrelation function (ACF) of the residuals:

```
> acf(m)                  # Plot the ACF of the model residuals
```

Discussion

The Durbin–Watson test is often used in time series analysis, but it was originally created for diagnosing autocorrelation in regression residuals. Autocorrelation in the residuals is a scourge because it distorts the regression statistics, such as the *F* statistic and the *t* statistics for the regression coefficients. The presence of autocorrelation suggests that your model is missing a useful predictor variable or that it should include a time series component, such as a trend or a seasonal indicator.

This first example builds a simple regression model and then tests the residuals for autocorrelation. The test returns a *p*-value of 0.07755, which indicates that there is no significant autocorrelation:

```
> m <- lm(y ~ x)
> dwtest(m)

        Durbin-Watson test

data:  m
DW = 1.5654, p-value = 0.07755
alternative hypothesis: true autocorrelation is greater than 0
```

This second example exhibits autocorrelation in the residuals. The *p*-value is only 0.01298, so the autocorrelation is likely positive:

```
> m <- lm(y ~ z)
> dwtest(m)

        Durbin-Watson test

data:  m
DW = 1.2946, p-value = 0.01298
alternative hypothesis: true autocorrelation is greater than 0
```

By default, `dwtest` performs a one-sided test and answers this question: Is the autocorrelation of the residuals greater than zero? If your model could exhibit negative autocorrelation (yes, that is possible), then you should use the `alternative` option to perform a two-sided test:

```
> dwtest(m, alternative="two.sided")
```

The Durbin–Watson test is also implemented by the `durbin.watson` function of the car package. I suggested the `dwtest` function only because I think the output is easier to read.

See Also

Neither the lmtest test package nor the car package are included in the standard distribution of R; see Recipes 3.6 and 3.9. See Recipes 14.13 and 14.16 for more regarding tests of autocorrelation.

11.18 Predicting New Values

Problem

You want to predict new values from your regression model.

Solution

Save the predictor data in a data frame. Use the `predict` function, setting the `newdata` parameter to the data frame:

```
> m <- lm(y ~ u + v + w)
> preds <- data.frame(u=3.1, v=4.0, w=5.5)
> predict(m, newdata=preds)
```

Discussion

Once you have a linear model, making predictions is quite easy because the `predict` function does all the heavy lifting. The only annoyance is arranging for a data frame to contain your data.

The `predict` function returns a vector of predicted values with one prediction for every row in the data. The example in the Solution contains one row, so `predict` returns one value:

```
> preds <- data.frame(u=3.1, v=4.0, w=5.5)
> predict(m, newdata=preds)
       1
12.31374
```

If your predictor data contains several rows, you get one prediction per row:

```
> preds <- data.frame(
+             u=c(3.0, 3.1, 3.2, 3.3),
```

```
+                    v=c(3.9, 4.0, 4.1, 4.2),
+                    w=c(5.3, 5.5, 5.7, 5.9) )
> predict(m, newdata=preds)
        1        2        3        4
 11.97277 12.31374 12.65472 12.99569
```

In case it's not obvious: the new data needn't contain values for response variables, only predictor variables. After all, you are trying to *calculate* the response, so it would be unreasonable of R to expect you to supply it.

See Also

These are just the point estimates of the predictions. See Recipe 11.19 for the confidence intervals.

11.19 Forming Prediction Intervals

Problem

You are making predictions using a linear regression model. You want to know the prediction intervals: the range of the distribution of the prediction.

Solution

Use the predict function and specify interval="prediction":

```
> predict(m, newdata=preds, interval="prediction")
```

Discussion

This is a continuation of Recipe 11.18, which described packaging your data into a data frame for the predict function. We are adding interval="prediction" to obtain prediction intervals.

Here is the example from Recipe 11.18, now with prediction intervals. The new lwr and upr columns are the lower and upper limits, respectively, for the interval:

```
> predict(m, newdata=preds, interval="prediction")
       fit       lwr      upr
1 11.97277 7.999848 15.94569
2 12.31374 8.272327 16.35516
3 12.65472 8.540045 16.76939
4 12.99569 8.803251 17.18813
```

By default, predict uses a confidence level of 0.95. You can change this via the level argument.

A word of caution: these prediction intervals are extremely sensitive to deviations from normality. If you suspect that your response variable is not normally distributed, consider a nonparametric technique, such as the bootstrap (Recipe 13.8), for prediction intervals.

11.20 Performing One-Way ANOVA

Problem

Your data is divided into groups, and the groups are normally distributed. You want to know if the groups have significantly different means.

Solution

Use a factor to define the groups. Then apply the oneway.test function:

```
> oneway.test(x ~ f)
```

Here, x is a vector of numeric values and f is a factor that identifies the groups. The output includes a p-value. Conventionally, a p-value of less than 0.05 indicates that two or more groups have significantly different means whereas a value exceeding 0.05 provides no such evidence.

Discussion

Comparing the means of groups is a common task. One-way ANOVA performs that comparison and computes the probability that they are statistically identical. A small p-value indicates that two or more groups likely have different means. (It does *not* indicate that *all* groups have different means.)

The basic ANOVA test assumes that your data has a normal distribution or that, at least, it is pretty close to bell-shaped. If not, use the Kruskal–Wallis test instead (Recipe 11.23).

We can illustrate ANOVA with stock market historical data. Is the stock market more profitable in some months than in others? For instance, a common folk myth says that October is a bad month for stock market investors.[‡] I explored this question by creating a vector and a factor. The vector, r, contains the daily percentage change in the Standard & Poor's 500 index, a broad measure of stock market performance. The factor, mon, indicates the calendar month in which that change occurred: Jan, Feb, Mar, and so forth. The data covers the period 1950 though 2009.

The one-way ANOVA shows a p-value of 0.03347:

```
> oneway.test(r ~ mon)

        One-way analysis of means (not assuming equal variances)

data:  r and mon
F = 1.9102, num df = 11.000, denom df = 5930.142, p-value = 0.03347
```

[‡] In the words of Mark Twain, "October: This is one of the peculiarly dangerous months to speculate in stocks in. The others are July, January, September, April, November, May, March, June, December, August and February."

We can conclude that stock market changes varied significantly according to the calendar month.

Before you run to your broker and start flipping your portfolio monthly, however, we should check something: did the pattern change recently? We can limit the analysis to recent data by specifying a subset parameter. This works for oneway.test just as it does for the lm function. The subset contains the indexes of observations to be analyzed; all other observations are ignored. Here, we give the indexes of the 2,500 most recent observations, which is about 10 years of data:

```
> oneway.test(r ~ mon, subset=tail(seq_along(r),2500))

        One-way analysis of means (not assuming equal variances)

data:  r and mon
F = 0.7973, num df = 11.000, denom df = 975.963, p-value = 0.643
```

Uh-oh! Those monthly differences evaporated during the past 10 years. The large *p*-value, 0.643, indicates that changes have not recently varied according to calendar month. Apparently, those differences are a thing of the past.

Notice that the oneway.test output says "(not assuming equal variances)". If you know the groups have equal variances, you'll get a less conservative test by specifying var.equal=TRUE:

```
> oneway.test(x ~ f, var.equal=TRUE)
```

You can also perform one-way ANOVA by using the aov function like this:

```
> m <- aov(x ~ f)
> summary(m)
```

However, the aov function always assumes equal variances and so is somewhat less flexible than oneway.test.

See Also

If the means are significantly different, use Recipe 11.22 to see the actual differences. Use Recipe 11.23 if your data is not normally distributed, as required by ANOVA.

11.21 Creating an Interaction Plot

Problem

You are performing multiway ANOVA: using two or more categorical variables as predictors. You want a visual check of possible interaction between the predictors.

Solution

Use the `interaction.plot` function:

```
> interaction.plot(pred1, pred2, resp)
```

Here, `pred1` and `pred2` are two categorical predictors and `resp` is the response variable.

Discussion

ANOVA is a form of linear regression, so ideally there is a linear relationship between every predictor and the response variable. One source of nonlinearity is an *interaction* between two predictors: as one predictor changes value, the other predictor changes its relationship to the response variable. Checking for interaction between predictors is a basic diagnostic.

The `faraway` package contains a dataset called `rats`. In it, `treat` and `poison` are categorical variables and `time` is the response variable. When plotting `poison` against `time` we are looking for straight, parallel lines, which indicate a linear relationship. However, using the `interaction.plot` function reveals that something is not right:

```
> library(faraway)
> data(rats)
> interaction.plot(rats$poison, rats$treat, rats$time)
```

The resulting graph is shown in Figure 11-6. Each line graphs `time` against `poison`. The difference between lines is that each line is for a different value of `treat`. The lines should be parallel, but the top two are not exactly parallel. Evidently, varying the value of `treat` "warped" the lines, introducing a nonlinearity into the relationship between `poison` and `time`.

This signals a possible interaction that we should check. For this data it just so happens that yes, there is an interaction but no, it is not statistically significant. The moral is clear: the visual check is useful, but it's not foolproof. Follow up with a statistical check.

See Also

See Recipe 11.6.

11.22 Finding Differences Between Means of Groups

Problem

Your data is divided into groups, and an ANOVA test indicates that the groups have significantly different means. You want to know the differences between those means for all groups.

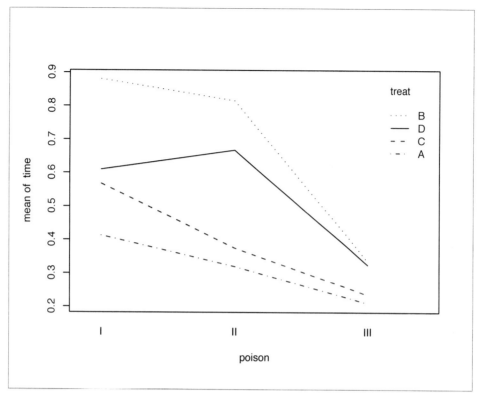

Figure 11-6. Plot of interaction between treat and poison

Solution

Perform the ANOVA test using the aov function, which returns a model object. Then apply the TukeyHSD function to the model object:

```
> m <- aov(x ~ f)
> TukeyHSD(m)
```

Here, x is your data and f is the grouping factor. You can plot the TukeyHSD result to obtain a graphical display of the differences:

```
> plot(TukeyHSD(m))
```

Discussion

The ANOVA test is important because it tells you whether or not the groups' means are different. But the test does not identify *which* groups are different, and it does not report their differences.

The TukeyHSD function can calculate those differences and help you identify the largest ones. It uses the "honest significant differences" method invented by John Tukey.

I'll illustrate TukeyHSD by continuing the example from Recipe 11.20, which grouped daily stock market changes by month. Here, I will group them by weekday instead, using a factor called wday that identifies the day of the week (Mon, ..., Fri) on which the change occurred. I'll use the first 2,500 observations, which roughly cover the period from 1950 to 1960:

```
> oneway.test(r ~ wday, subset=1:2500)

        One-way analysis of means (not assuming equal variances)

data:  r and wday
F = 12.6051, num df = 4.000, denom df = 1242.499, p-value = 4.672e-10
```

The p-value is essentially zero, indicating that average changes varied significantly depending on the weekday. To use the TukeyHSD function, I first perform the ANOVA test using the aov function, which returns a model object, and then apply the TukeyHSD function to the object:

```
> m <- aov(r ~ wday, subset=1:2500)
> TukeyHSD(m)
  Tukey multiple comparisons of means
    95% family-wise confidence level

Fit: aov(formula = r ~ wday, subset = 1:2500)

$wday
               diff          lwr       upr     p adj
Tue-Mon  0.13131281  0.007025859 0.2555998 0.0323054
Wed-Mon  0.23846814  0.114846431 0.3620898 0.0000015
Thu-Mon  0.22371916  0.099432205 0.3480061 0.0000094
Fri-Mon  0.31728974  0.192816883 0.4417626 0.0000000
Wed-Tue  0.10715532 -0.015961497 0.2302721 0.1222470
Thu-Tue  0.09240635 -0.031378436 0.2161911 0.2481996
Fri-Tue  0.18597693  0.062005489 0.3099484 0.0004185
Thu-Wed -0.01474898 -0.137865796 0.1083678 0.9975329
Fri-Wed  0.07882161 -0.044482882 0.2021261 0.4064721
Fri-Thu  0.09357058 -0.030400857 0.2175420 0.2378246
```

Each line in the output table includes the difference between the means of two groups (diff) as well as the lower and upper bounds of the confidence interval (lwr and upr) for the difference. The first line in the table, for example, compares the Tue group and the Wed group: the difference of their means is 0.1313 with a confidence interval of (0.0070, 0.2556).

Scanning the table, we see that the Fri–Mon comparison had the largest difference, which was 0.3173.

A cool feature of TukeyHSD is that it can display these differences visually, too. Simply plot the function's return value:

```
> plot(TukeyHSD(m))
```

Figure 11-7 shows the graph for our stock market example. The horizontal lines plot the confidence intervals for each pair. With this visual representation you can quickly see that several confidence intervals cross over zero, indicating that the difference is not necessarily significant. You can also see that the Fri–Mon pair has the largest difference because their confidence interval is farthest to the right.

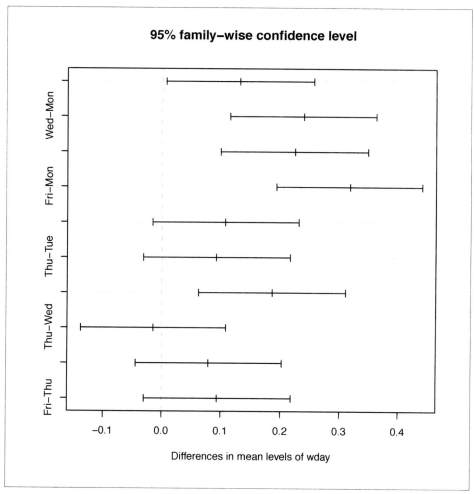

Figure 11-7. Differences between group means

See Also

See Recipe 11.20.

11.23 Performing Robust ANOVA (Kruskal–Wallis Test)

Problem

Your data is divided into groups. The groups are not normally distributed, but their distributions have similar shapes. You want to perform a test similar to ANOVA—you want to know if the group medians are significantly different.

Solution

Create a factor that defines the groups of your data. Use the kruskal.test function, which implements the Kruskal–Wallis test. Unlike the ANOVA test, this test does not depend upon the normality of the data:

```
> kruskal.test(x ~ f)
```

Here, x is a vector of data and f is a grouping factor. The output includes a p-value. Conventionally, $p < 0.05$ indicates that there is a significant difference between the medians of two or more groups whereas $p > 0.05$ provides no such evidence.

Discussion

Regular ANOVA assumes that your data has a Normal distribution. It can tolerate some deviation from normality, but extreme deviations will produce meaningless p-values.

The Kruskal–Wallis test is a nonparametric version of ANOVA, which means that it does not assume normality. However, it does assume same-shaped distributions. You should use the Kruskal–Wallis test whenever your data distribution is nonnormal or simply unknown.

The null hypothesis is that all groups have the same median. Rejecting the null hypothesis (with $p < 0.05$) does not indicate that *all* groups are different, but it does suggest that two or more groups are different.

One year, I taught Business Statistics to 94 undergraduate students. The class included a midterm examination, and there were four homework assignments prior to the exam. I wanted to know: What is the relationship between completing the homework and doing well on the exam? If there is no relation, then the homework is irrelevant and needs rethinking.

I created a vector of grades, one per student. I also created a parallel factor that captured the number of homework assignments completed by that student:

```
> print(midterm)
 [1] 0.8182 0.6818 0.5114 0.6705 0.6818 0.9545 0.9091 0.9659 1.0000 0.7273
[11] 0.7727 0.7273 0.6364 0.6250 0.6932 0.6932 0.0000 0.8750 0.8068 0.9432
.
. (etc.)
.
[91] 0.9006 0.7029 0.9171 0.9628
```

```
> print(hw)
 [1] 4 4 2 3 4 4 4 3 4 4 3 3 1 4 4 3 3 4 3 4 4 4 3 2 4 4 4 4 3 2 3 4 4 0 4 4 4 4
[39] 4 4 4 4 4 4 4 4 4 4 4 4 4 4 4 4 4 4 2 4 4 4 4 4 1 4 4 4 4 4 4 4 4 4 2 4 4
[77] 3 4 2 2 4 4 4 4 1 3 0 4 4 0 4 3 4 4
Levels: 0 1 2 3 4
```

Notice that the hw variable—although it appears to be numeric—is actually a factor. It assigns each midterm grade to one of five groups depending upon how many homework assignments the student completed.

The distribution of exam grades is definitely not Normal: the students have a wide range of math skills, so there are an unusual number of A and F grades. Hence regular ANOVA would not be appropriate. Instead I used the Kruskal–Wallis test and obtained a *p*-value of essentially zero (3.99×10^{-5}, or 0.00003669):

```
> kruskal.test(midterm ~ hw)

        Kruskal-Wallis rank sum test

data:  midterm by hw
Kruskal-Wallis chi-squared = 25.6813, df = 4, p-value = 3.669e-05
```

Obviously, there is a significant performance difference between students who complete their homework and those who do not. But what could I actually conclude? At first, I was pleased that the homework appeared so effective. Then it dawned on me that this was a classic error in statistical reasoning: I assumed that correlation implied causality. It does not, of course. Perhaps strongly motivated students do well on both homework and exams whereas lazy students do not. In that case, the causal factor is degree of motivation, not the brilliance of my homework selection. In the end, I could only conclude something very simple: students who complete the homework will likely do well on the midterm exam, but I don't really know why.

11.24 Comparing Models by Using ANOVA

Problem

You have two models of the same data, and you want to know whether they produce different results.

Solution

The anova function can compare two models and report if they are significantly different:

```
> anova(m1, m2)
```

Here, m1 and m2 are both model objects returned by lm. The output from anova includes a *p*-value. Conventionally, a *p*-value of less than 0.05 indicates that the models are significantly different whereas a value exceeding 0.05 provides no such evidence.

Discussion

In Recipe 11.3, we used the anova function to print the ANOVA table for one regression model. Now we are using the two-argument form to compare two models.

The anova function has one strong requirement when comparing two models: one model must be contained within the other. That is, all the terms of the smaller model must appear in the larger model. Otherwise, the comparison is impossible.

The ANOVA analysis performs an F test that is similar to the F test for a linear regression. The difference is that this test is between two models whereas the regression F test is between using the regression model and using no model.

Suppose we build three models of y, adding terms as we go:

```
> m1 <- lm(y ~ u)
> m2 <- lm(y ~ u + v)
> m3 <- lm(y ~ u + v + w)
```

Is m2 really different from m1? We can use anova to compare them, and the result is a p-value of 0.003587:

```
> anova(m1, m2)
Analysis of Variance Table

Model 1: y ~ u
Model 2: y ~ u + v
  Res.Df     RSS Df Sum of Sq     F    Pr(>F)
1     28 108.968
2     27  79.138  1     29.83 10.177 0.003587 **
---
Signif. codes:  0 '***' 0.001 '**' 0.01 '*' 0.05 '.' 0.1 ' ' 1
```

The small p-value indicates that the models are significantly different. Comparing m2 and m3, however, yields a p-value of 0.05744:

```
> anova(m2, m3)
Analysis of Variance Table

Model 1: y ~ u + v
Model 2: y ~ u + v + w
  Res.Df    RSS Df Sum of Sq      F  Pr(>F)
1     27 79.138
2     26 68.696  1    10.442 3.9522 0.05744 .
---
Signif. codes:  0 '***' 0.001 '**' 0.01 '*' 0.05 '.' 0.1 ' ' 1
```

This is right on the edge. Strictly speaking, it does not pass our requirement to be smaller than 0.05; however, it's close enough that you might judge the models to be "different enough."

This example is a bit contrived, so it does not show the larger power of anova. I use anova when, while experimenting with complicated models by adding and deleting multiple terms, I need to know whether or not the new model is really different from the original one. In other words: if I add terms and the new model is essentially unchanged, then the extra terms are not worth the additional complications.

Useful Tricks

Introduction

The recipes in this chapter are neither obscure numerical calculations nor deep statistical techniques. Yet they are useful functions and idioms that you will likely need at one time or another.

12.1 Peeking at Your Data

Problem

You have a lot of data—too much to display at once. Nonetheless, you want to see some of the data.

Solution

Use head to view the first few data or rows:

```
> head(x)
```

Use tail to view the last few data or rows:

```
> tail(x)
```

Discussion

Printing a large dataset is pointless because everything just rolls off your screen. Use head to see a little bit of the data:

```
> head(dfrm)
          x          y          z
1  0.7533110 0.57562846 -0.1710760
2  2.0143547 0.83312274  0.3698584
3 -0.3551345 0.57471542  2.0132348
4  2.0281678 0.78945319 -0.5378854
```

```
5 -2.2168745  0.01758024   1.8344879
6  0.7583962 -1.78214755   2.2848990
```

Use `tail` to see the last few rows and the number of rows. Here, we see that this data frame has 10,120 rows:

```
> tail(dfrm)
              x            y          z
10115 -0.0314354 -0.74988291 -0.2048963
10116 -0.4779001  0.93407510  1.0509977
10117 -1.1314402  1.89308417  1.7207972
10118  0.4891881 -1.20792811 -1.4630227
10119  1.2349013 -0.09615198 -0.9887513
10120 -1.3763834 -2.25309628  0.9296106
```

See Also

See Recipe 12.15 for seeing the structure of your variable's contents.

12.2 Widen Your Output

Problem

You are printing wide datasets. R is wrapping the output, making it hard to read.

Solution

Set the `width` option to reflect the true number of columns in your output window:

```
> options(width=numcols)
```

Discussion

By default, R formats your output to be 80 columns wide. It's frustrating when you print a wide dataset and R wraps the output lines according to its assumed limitation of 80 columns. Your expensive, wide-screen display can probably accommodate many more columns than that.

If your display is 120 characters wide, for example, then ask R to use the entire 120 columns by setting the `width` option:

```
> options(width=120)
```

Now R will wrap its lines after 120 characters, not 80, letting you comfortably view wider datasets.

See Also

See Recipe 3.16 for more about setting options.

12.3 Printing the Result of an Assignment

Problem

You are assigning a value to a variable and you want to see its value.

Solution

Simply put parentheses around the assignment:

```
> x <- 1/pi          # Prints nothing
> (x <- 1/pi)        # Prints assigned value
[1] 0.3183099
```

Discussion

Normally, R inhibits printing when it sees you enter a simple assignment. When you surround the assignment with parentheses, however, it is no longer a simple assignment and so R prints the value.

See Also

See Recipe 2.1 for more ways to print things.

12.4 Summing Rows and Columns

Problem

You want to sum the rows or columns of a matrix or data frame.

Solution

Use rowSums to sum the rows:

```
> rowSums(m)
```

Use colSums to sum the columns:

```
> colSums(m)
```

Discussion

This is a mundane recipe, but it's so common that it deserves mentioning. I use this recipe, for example, when producing reports that include column totals. In this example, daily.prod is a record of this week's factory production and we want totals by product and by day:

```
> daily.prod
      Widgets Gadgets Thingys
Mon       179     167     182
```

```
Tue     153     193     166
Wed     183     190     170
Thu     153     161     171
Fri     154     181     186
> colSums(daily.prod)
Widgets Gadgets Thingys
    822     892     875
> rowSums(daily.prod)
Mon Tue Wed Thu Fri
528 512 543 485 521
```

These functions return a vector. In the case of column sums, we can append the vector to the matrix and thereby neatly print the data and totals together:

```
> rbind(daily.prod, Totals=colSums(daily.prod))
        Widgets Gadgets Thingys
Mon         179     167     182
Tue         153     193     166
Wed         183     190     170
Thu         153     161     171
Fri         154     181     186
Totals      822     892     875
```

12.5 Printing Data in Columns

Problem

You have several parallel data vectors, and you want to print them in columns.

Solution

Use cbind to form the data into columns; then print the result.

Discussion

When you have parallel vectors, it's difficult to see their relationship if you print them separately:

```
> print(x)
 [1] -0.3142317  2.4033751 -0.7182640 -1.7606110 -1.1252812 -0.7195406  1.3102240
 [8]  0.4518985  0.1521774  0.6533708
> print(y)
 [1] -0.9473355 -1.0710863 -0.1760594  1.8720502 -0.5103384  1.4388904  1.1531710
 [8] -0.1548398 -0.2599055  0.4713668
```

Use the cbind function to form them into columns that, when printed, show the data's structure:

```
> print(cbind(x,y))
              x          y
[1,] -0.3142317 -0.9473355
[2,]  2.4033751 -1.0710863
[3,] -0.7182640 -0.1760594
```

```
[4,]  -1.7606110   1.8720502
[5,]  -1.1252812  -0.5103384
[6,]  -0.7195406   1.4388904
[7,]   1.3102240   1.1531710
[8,]   0.4518985  -0.1548398
[9,]   0.1521774  -0.2599055
[10,]   0.6533708   0.4713668
```

You can include expressions in the output, too. Use a tag to give them a column heading:

```
> print(cbind(x,y,Total=x+y))
              x          y      Total
[1,]  -0.3142317  -0.9473355  -1.2615672
[2,]   2.4033751  -1.0710863   1.3322888
[3,]  -0.7182640  -0.1760594  -0.8943233
[4,]  -1.7606110   1.8720502   0.1114392
[5,]  -1.1252812  -0.5103384  -1.6356196
[6,]  -0.7195406   1.4388904   0.7193499
[7,]   1.3102240   1.1531710   2.4633949
[8,]   0.4518985  -0.1548398   0.2970587
[9,]   0.1521774  -0.2599055  -0.1077281
[10,]   0.6533708   0.4713668   1.1247376
```

12.6 Binning Your Data

Problem

You have a vector, and you want to split the data into groups according to intervals. Statisticians call this *binning* your data.

Solution

Use the cut function. You must define a vector, say breaks, which gives the ranges of the intervals. The cut function will group your data according to those intervals. It returns a factor whose levels (elements) identify each datum's group:

```
> f <- cut(x, breaks)
```

Discussion

This example generates 1,000 random numbers that have a standard Normal distribution. It breaks them into six groups by defining intervals at ±1, ±2, and ±3 standard deviations:

```
> x <- rnorm(1000)
> breaks <- c(-3,-2,-1,0,1,2,3)
> f <- cut(x, breaks)
```

The result is a factor, f, that identifies the groups. The summary function shows the number of elements by level. R creates names for each level, using the mathematical notation for an interval:

```
> summary(f)
(-3,-2] (-2,-1] (-1,0]   (0,1]   (1,2]   (2,3]   NA's
     18     119     348     355     137      21      2
```

The results are bell-shaped, which is what we expect. There are two NA values, indicating that two values in x fell outside the defined intervals.

We can use the `labels` parameter to give nice, predefined names to the six groups instead of the funky synthesized names:

```
> f <- cut(x, breaks, labels=c("Bottom", "Low", "Neg", "Pos", "High", "Top"))
```

Now the `summary` function uses our names:

```
Bottom    Low    Neg    Pos   High    Top   NA's
    18    119    348    355    137     21      2
```

Don't Throw Away That Information

Binning is useful for summaries such as histograms. But it results in *information loss*, which can be harmful in modeling. Consider the extreme case of binning a continuous variable into two values, "high" and "low". The binned data has only two possible values, so you have replaced a rich source of information with *one bit* of information. Where the continuous variable might be a powerful predictor, the binned variable can distinguish at most two states and so will likely have only a fraction of the original power. Before binning, I suggest exploring other transformations that are less lossy.

12.7 Finding the Position of a Particular Value

Problem

You have a vector. You know a particular value occurs in the contents, and you want to know its position.

Solution

The `match` function will search a vector for a particular value and return the position:

```
> vec <- c(100,90,80,70,60,50,40,30,20,10)
> match(80, vec)
[1] 3
```

Here `match` returns 3, which is the position of 80 within `vec`.

Discussion

There are special functions for finding the location of the minimum and maximum values—`which.min` and `which.max`, respectively:

```
> vec <- c(100,90,80,70,60,50,40,30,20,10)
> which.min(vec)          # Position of smallest element
```

```
[1] 10
> which.max(vec)          # Position of largest element
[1] 1
```

See Also

This technique is used in Recipe 11.12.

12.8 Selecting Every nth Element of a Vector

Problem

You want to select every *n*th element of a vector.

Solution

Create a logical indexing vector that is TRUE for every *n*th element. One approach is to find all subscripts that equal zero when taken modulo *n*:

```
> v[ seq_along(v) %% n == 0 ]
```

Discussion

This problem arises in systematic sampling: we want to sample a dataset by selecting every *n*th element. The seq_along(v) function generates the sequence of integers that can index v; it is equivalent to 1:length(v). We compute each index value modulo *n* by the expression:

```
seq_along(v) %% n
```

Then we find those values that equal zero:

```
seq_along(v) %% n == 0
```

The result is a logical vector, the same length as v and with TRUE at every *n*th element, that can index v to select the desired elements:

```
v[ seq_along(v) %% n == 0 ]
```

If you just want something simple like every second element, you can use the recycling rule in a clever way. Index v with a two-element logical vector, like this:

```
v[ c(FALSE, TRUE) ]
```

If v has more than two elements then the indexing vector is too short. Hence R will invoke the Recycling Rule and expand the index vector to the length of v, recycling its contents. That gives an index vector that is FALSE, TRUE, FALSE, TRUE, FALSE, TRUE, and so forth. Voilà! The final result is every second element of v.

See Also

See Recipe 5.3 for more about the Recycling Rule.

12.9 Finding Pairwise Minimums or Maximums

Problem

You have two vectors, v and w, and you want to find the minimums or the maximums of pairwise elements. That is, you want to calculate:

$$\min(v_1, w_1), \min(v_2, w_2), \min(v_3, w_3), ...$$

or:

$$\max(v_1, w_1), \max(v_2, w_2), \max(v_3, w_3), ...$$

Solution

R calls these the *parallel minimum* and the *parallel maximum*. The calculation is performed by pmin(v,w) and pmax(v,w), respectively:

```
> pmin(1:5, 5:1)        # Find the element-by-element minimum
[1] 1 2 3 2 1
> pmax(1:5, 5:1)        # Find the element-by-element maximum
[1] 5 4 3 4 5
```

Discussion

When an R beginner wants pairwise minimums or maximums, a common mistake is to write min(v,w) or max(v,w). Those are not pairwise operations: min(v,w) returns a single value, the minimum over all v and w. Likewise, max(v,w) returns a single value from all of v and w.

The pmin and pmax values compare their arguments in parallel, picking the minimum or maximum for each subscript. They return a vector that matches the length of the inputs.

You can combine pmin and pmax with the Recycling Rule to perform useful hacks. Suppose the vector v contains both positive and negative values, and you want to reset the negative values to zero. This does the trick:

```
> v <- pmax(v, 0)
```

By the Recycling Rule, R expands the zero-valued scalar into a vector of zeros that is the same length as v. Then pmax does an element-by-element comparison, taking the larger of zero and each element of v:

```
> v <- c(1, -2, 3, -4, 5, -6, 7)
> pmax(v,0)
[1] 1 0 3 0 5 0 7
```

Actually, pmin and pmax are more powerful than the Solution indicates. They can take more than two vectors, comparing all vectors in parallel.

See Also

See Recipe 5.3 for more about the Recycling Rule.

12.10 Generating All Combinations of Several Factors

Problem

You have two or more factors. You want to generate all combinations of their levels, also known as their Cartesian product.

Solution

Use the expand.grid function. Here, f and g are two factors:

```
> expand.grid(f, g)
```

Discussion

This code snippet creates two factors—sides represents the two sides of a coin, and faces represents the six faces of a die (those little spots on a die are called *pips*):

```
> sides <- factor(c("Heads", "Tails"))
> faces <- factor(c("1 pip", paste(2:6, "pips")))
```

We can use expand.grid to find all combinations of one roll of the die and one coin toss:

```
> expand.grid(faces, sides)
      Var1  Var2
1   1 pip Heads
2  2 pips Heads
3  3 pips Heads
4  4 pips Heads
5  5 pips Heads
6  6 pips Heads
7   1 pip Tails
8  2 pips Tails
9  3 pips Tails
10 4 pips Tails
11 5 pips Tails
12 6 pips Tails
```

Similarly, we can find all combinations of two dice:

```
> expand.grid(faces, faces)
      Var1   Var2
1   1 pip  1 pip
2  2 pips  1 pip
3  3 pips  1 pip
4  4 pips  1 pip
5  5 pips  1 pip
6  6 pips  1 pip
7   1 pip 2 pips
```

```
8  2 pips 2 pips
.
. (etc.)
.
34 4 pips 6 pips
35 5 pips 6 pips
36 6 pips 6 pips
```

The result is a data frame. R automatically provides the row names and column names.

The Solution and the example show the Cartesian product of two factors, but expand.grid can handle three or more factors, too.

See Also

If you're working with strings and not factors, then you can also use Recipe 7.7 to generate combinations.

12.11 Flatten a Data Frame

Problem

You have a data frame of numeric values. You want to process all its elements together, not as separate columns—for example, to find the mean across all values.

Solution

Convert the data frame to a matrix and then process the matrix. This example finds the mean of all elements in the data frame dfrm:

```
> mean(as.matrix(dfrm))
```

It is sometimes necessary then to convert the matrix to a vector. In that case, use as.vector(as.matrix(dfrm)).

Discussion

Suppose we have a data frame, such as the factory production data from Recipe 12.4:

```
> daily.prod
    Widgets Gadgets Thingys
Mon     179     167     182
Tue     153     193     166
Wed     183     190     170
Thu     153     161     171
Fri     154     181     186
```

Suppose also that we want the average daily production across all days and products. This won't work:

```
> mean(daily.prod)
Widgets Gadgets Thingys
  164.4   178.4   175.
```

The mean function thinks it should average individual columns, which is usually what you want. But when you want the average across all values, first collapse the data frame down to a matrix:

```
> mean(as.matrix(daily.prod))
[1] 172.6
```

This recipe works only on data frames with all-numeric data. Recall that converting a data frame with mixed data (numeric columns mixed with character columns or factors) into a matrix forces all columns to be converted to characters.

See Also

See Recipe 5.33 for more about converting between datatypes.

12.12 Sorting a Data Frame

Problem

You have a data frame. You want to sort the contents, using one column as the sort key.

Solution

Use the order function on the sort key, and then rearrange the data frame rows according to that ordering:

```
> dfrm <- dfrm[order(dfrm$key),]
```

Here dfrm is a data frame and dfrm$key is the sort-key column.

Discussion

The sort function is great for vectors but is ineffective for data frames. Sorting a data frame requires an extra step. Suppose we have the following data frame and we want to sort by month:

```
> print(dfrm)
  month day outcome
1     7  11     Win
2     8  10    Lose
3     8  25     Tie
4     6  27     Tie
5     7  22     Win
```

The order function tells us how to rearrange the months into ascending order:

```
> order(dfrm$month)
[1] 4 1 5 2 3
```

It does not return data values; rather, it returns the subscripts of dfrm$month that would sort it. We use those subscripts to rearrange the entire data frame:

```
> dfrm[order(dfrm$month),]
  month day outcome
4     6  27     Tie
1     7  11     Win
5     7  22     Win
2     8  10    Lose
3     8  25     Tie
```

After rearranging the data frame, the month column is in ascending order—just as we wanted.

12.13 Sorting by Two Columns

Problem

You want to sort the contents of a data frame, using two columns as sort keys.

Solution

Sorting by two columns is similar to sorting by one column. The order function accepts multiple arguments, so we can give it two sort keys:

```
> dfrm <- dfrm[order(dfrm$key1,dfrm$key2),]
```

Discussion

In Recipe 12.12 we used the order function to rearrange data frame columns. We can give a second sort key to order, and it will use the additional key to break ties in the first key.

Continuing our example from Recipe 12.12, we can sort on both month and day by giving both those columns to order:

```
> dfrm[order(dfrm$month,dfrm$day),]
  month day outcome
4     6  27     Tie
1     7  11     Win
5     7  22     Win
2     8  10    Lose
3     8  25     Tie
```

Within months 7 and 8, the days are now sorted into ascending order.

See Also

See Recipe 12.12.

12.14 Stripping Attributes from a Variable

Problem

A variable is carrying around old attributes. You want to remove some or all of them.

Solution

To remove all attributes, assign NULL to the variable's `attributes` property:

```
> attributes(x) <- NULL
```

To remove a single attribute, select the attribute using the `attr` function, and set it to NULL:

```
> attr(x, "attributeName") <- NULL
```

Discussion

Any variable in R can have attributes. An *attribute* is simply a name/value pair, and the variable can have many of them. A common example is the dimensions of a matrix variable, which are stored in an attribute. The attribute name is `dim` and the attribute value is a two-element vector giving the number of rows and columns.

You can view the attributes of x by printing `attributes(x)` or `str(x)`.

Sometimes you want just a number and R insists on giving it attributes. This can happen when you fit a simple linear model and extract the slope, which is the second regression coefficient:

```
> m <- lm(resp ~ pred)
> slope <- coef(m)[2]
> slope
    pred
1.011960
```

When we print `slope`, R also prints `"pred"`. That is a name attribute given by `lm` to the coefficient (because it's the coefficient for the `pred` variable). We can see that more clearly by printing the internals of `slope`, which reveals a `"names"` attribute:

```
> str(slope)
 Named num 1.01
 - attr(*, "names")= chr "pred"
```

Stripping off all the attributes is easy, after which the slope value becomes simply a number:

```
> attributes(slope) <- NULL        # Strip off all attributes
> str(slope)                       # Now the "names" attribute is gone
 num 1.01
> slope                            # And the number prints cleanly without a label
[1] 1.011960
```

Alternatively, we could have stripped the single offending attribute this way:

```
> attr(slope, "names") <- NULL
```

 Remember that a matrix is a vector (or list) with a `dim` attribute. If you strip all the attributes from a matrix, that will strip away the dimensions and thereby turn it into a mere vector (or list). Furthermore, stripping the attributes from an object (specifically, an S3 object) can render it useless. So remove attributes with care.

See Also

See Recipe 12.15 for more about seeing attributes.

12.15 Revealing the Structure of an Object

Problem

You called a function that returned something. Now you want to look inside that something and learn more about it.

Solution

Use `class` to determine the thing's object class:

```
> class(x)
```

Use `mode` to strip away the object-oriented features and reveal the underlying structure:

```
> mode(x)
```

Use `str` to show the internal structure and contents:

```
> str(x)
```

Discussion

I am amazed how often I call a function, get something back, and wonder: "What the heck is this thing?" Theoretically, the function documentation should explain the returned value, but somehow I feel better when I can see its structure and contents myself. This is especially true for objects with a nested structure: objects within objects.

Let's dissect the value returned by `lm` (the linear modeling function) in the simplest linear regression recipe, Recipe 11.1:

```
> m <- lm(y ~ x)
> print(m)

Call:
lm(formula = y ~ x)

Coefficients:
```

```
(Intercept)          x
    17.72          3.25
```

Always start by checking the thing's class. The class indicates if it's a vector, matrix, list, data frame, or object:

```
> class(m)
[1] "lm"
```

Hmmm. It seems that m is an object of class lm. That doesn't mean anything to me. I know, however, that all object classes are built upon the native data structures (vector, matrix, list, or data frame), so I use mode to strip away the object facade and reveal the underlying structure:

```
> mode(m)
[1] "list"
```

Ah-ha! It seems that m is built on a list structure. Now I can use list functions and operators to dig into its contents. First, I want to know the names of its list elements:

```
> names(m)
 [1] "coefficients"  "residuals"     "effects"       "rank"        "fitted.values"
 [6] "assign"        "qr"            "df.residual"   "xlevels"     "call"
[11] "terms"         "model"
```

The first list element is called "coefficients". I'm guessing those are the regression coefficients. Let's have a look:

```
> m$coefficients
(Intercept)          x
  17.722623    3.249677
```

Yes, that's what they are. I recognize those values.

I could continue digging into the list structure of m, but that would get tedious. The str function does a good job of revealing the internal structure of any variable. The dump of m is shown in Example 12-1.

Example 12-1. Dumping the structure of a variable

```
> str(m)
List of 12
 $ coefficients : Named num [1:2] 17.72 3.25
  ..- attr(*, "names")= chr [1:2] "(Intercept)" "x"
 $ residuals    : Named num [1:30] -12.471 -3.983 -2.899 0.192 14.302 ...
  ..- attr(*, "names")= chr [1:30] "1" "2" "3" "4" ...
 $ effects      : Named num [1:30] -372.443 155.463 -0.614 2.418 16.552 ...
  ..- attr(*, "names")= chr [1:30] "(Intercept)" "x" "" "" ...
 $ rank         : int 2
 $ fitted.values: Named num [1:30] 17.9 23.9 26.8 32.2 30 ...
  ..- attr(*, "names")= chr [1:30] "1" "2" "3" "4" ...
 $ assign       : int [1:2] 0 1
 $ qr           :List of 5
  ..$ qr   : num [1:30, 1:2] -5.477 0.183 0.183 0.183 0.183 ...
  .. ..- attr(*, "dimnames")=List of 2
  .. .. ..$ : chr [1:30] "1" "2" "3" "4" ...
```

```
.. .. ..$ : chr [1:2] "(Intercept)" "x"
.. ..- attr(*, "assign")= int [1:2] 0 1
..$ qraux: num [1:2] 1.18 1.23
..$ pivot: int [1:2] 1 2
..$ tol  : num 1e-07
..$ rank : int 2
..- attr(*, "class")= chr "qr"
$ df.residual : int 28
$ xlevels     : list()
$ call        : language lm(formula = y ~ x)
$ terms       :Classes 'terms', 'formula' length 3 y ~ x
.. ..- attr(*, "variables")= language list(y, x)
.. ..- attr(*, "factors")= int [1:2, 1] 0 1
.. .. ..- attr(*, "dimnames")=List of 2
.. .. .. ..$ : chr [1:2] "y" "x"
.. .. .. ..$ : chr "x"
.. ..- attr(*, "term.labels")= chr "x"
.. ..- attr(*, "order")= int 1
.. ..- attr(*, "intercept")= int 1
.. ..- attr(*, "response")= int 1
.. ..- attr(*, ".Environment")=<environment: R_GlobalEnv>
.. ..- attr(*, "predvars")= language list(y, x)
.. ..- attr(*, "dataClasses")= Named chr [1:2] "numeric" "numeric"
.. .. ..- attr(*, "names")= chr [1:2] "y" "x"
$ model       :'data.frame': 30 obs. of  2 variables:
..$ y: num [1:30] 5.41 19.94 23.92 32.43 44.26 ...
..$ x: num [1:30] 0.0478 1.9086 2.7999 4.4676 3.7649 ...
..- attr(*, "terms")=Classes 'terms', 'formula' length 3 y ~ x
.. .. ..- attr(*, "variables")= language list(y, x)
.. .. ..- attr(*, "factors")= int [1:2, 1] 0 1
.. .. .. ..- attr(*, "dimnames")=List of 2
.. .. .. .. ..$ : chr [1:2] "y" "x"
.. .. .. .. ..$ : chr "x"
.. .. ..- attr(*, "term.labels")= chr "x"
.. .. ..- attr(*, "order")= int 1
.. .. ..- attr(*, "intercept")= int 1
.. .. ..- attr(*, "response")= int 1
.. .. ..- attr(*, ".Environment")=<environment: R_GlobalEnv>
.. .. ..- attr(*, "predvars")= language list(y, x)
.. .. ..- attr(*, "dataClasses")= Named chr [1:2] "numeric" "numeric"
.. .. .. ..- attr(*, "names")= chr [1:2] "y" "x"
- attr(*, "class")= chr "lm"
```

Notice that str shows all the elements of m and then recursively dumps each element's contents and attributes. Long vectors and lists are truncated to keep the output manageable.

There is an art to exploring an R object. Use class, mode, and str to dig through the layers.

12.16 Timing Your Code

Problem

You want to know how much time is required to run your code. This is useful, for example, when you are optimizing your code and need "before" and "after" numbers to measure the improvement.

Solution

The `system.time` function will execute your code and then report the execution time:

```
> system.time(aLongRunningExpression)
```

The output is three to five numbers, depending on your platform. The first three are usually the most important:

User CPU time
: Number of CPU seconds spent running R

System CPU time
: Number of CPU seconds spent running the operating system, typically for input/output

Elapsed time
: Number of seconds on the clock

Discussion

Suppose we want to know the time required to generate 10,000,000 random normal variates and sum them:

```
> system.time(sum(rnorm(10000000)))
[1] 3.06 0.01 3.14   NA   NA
```

The output from `system.time` shows that R used 3.06 seconds of CPU time; the operating system used 0.01 seconds of CPU time; and 3.14 seconds elapsed during the test. (The NA values are subprocess times; in this example, there were no subprocesses.)

The third value, elapsed time, is not simply the sum of User CPU time and System CPU time. If your computer is busy running other processes, then more clock time will elapse while R is sharing the CPU with those other processes.

12.17 Suppressing Warnings and Error Messages

Problem

A function is producing annoying error messages or warning messages. You don't want to see them.

Solution

Surround the function call with suppressMessage(...) or suppressWarnings(...):

```
> suppressMessage(annoyingFunction())
> suppressuWarnings(annoyingFunction())
```

Discussion

I use the adf.test function quite a bit. However, it produces an annoying warning message, shown here at the bottom of the output, when the *p*-value is below 0.01:

```
> library(tseries)
> adf.test(x)

        Augmented Dickey-Fuller Test

data:  x
Dickey-Fuller = -4.3188, Lag order = 4, p-value = 0.01
alternative hypothesis: stationary

Warning message:
In adf.test(x) : p-value smaller than printed p-value
```

Fortunately, I can muzzle the function by calling it inside suppressWarnings(...):

```
> suppressWarnings(adf.test(x))

        Augmented Dickey-Fuller Test

data:  x
Dickey-Fuller = -4.3188, Lag order = 4, p-value = 0.01
alternative hypothesis: stationary
```

Notice that the warning message disappeared. The message is not entirely lost because R retains it internally. I can retrieve the message at my leisure by using the warnings function:

```
> warnings()
Warning message:
In adf.test(x) : p-value smaller than printed p-value
```

Some functions also produce "messages" (in R terminology), which are even more benign than warnings. Typically, they are merely informative and not signals of problems. If such a message is annoying you, call the function inside suppressMes sages(...), and the message will disappear.

See Also

See the options function for other ways to control the reporting of errors and warnings.

12.18 Taking Function Arguments from a List

Problem

Your data is captured in a list structure. You want to pass the data to a function, but the function does not accept a list.

Solution

In simple cases, convert the list to a vector. For more complex cases, the do.call function can break the list into individual arguments and call your function:

```
> do.call(function, list)
```

Discussion

If your data is captured in a vector, life is simple and most R functions work as expected:

```
> vec <- c(1, 3, 5, 7, 9)
> mean(vec)
[1] 5
```

If your data is captured in a list, then some functions complain and return a useless result, like this:

```
> numbers <- list(1, 3, 5, 7, 9)
> mean(numbers)
[1] NA
Warning message:
In mean.default(numbers) : argument is not numeric or logical: returning NA
```

The numbers list is a simple, one-level list, so we can just convert it to a vector and call the function:

```
> mean(unlist(numbers))
[1] 5
```

The big headaches come when you have *multilevel* list structures: lists within lists. These can occur within complex data structures. Here is a list of lists in which each sublist is a column of data:

```
> lists <- list(col1=list(7,8,9), col2=list(70,80,90), col3=list(700,800,900))
```

Suppose we want to form this data into a matrix. The cbind function is supposed to create data columns, but it gets confused by the list structure and returns something useless:

```
> cbind(lists)
     lists
col1 List,3
col2 List,3
col3 List,3
```

If we unlist the data then we just get one big, long column:

```
> cbind(unlist(lists))
       [,1]
col11    7
col12    8
col13    9
col21   70
col22   80
col23   90
col31  700
col32  800
col33  900
```

The solution is to use do.call, which splits the list into individual items and then calls cbind on those items:

```
> do.call(cbind, lists)
     col1 col2 col3
[1,] 7    70   700
[2,] 8    80   800
[3,] 9    90   900
```

Using do.call in that way is functionally identical to calling cbind like this:

```
> cbind(lists[[1]], lists[[2]], lists[[3]])
     [,1] [,2] [,3]
[1,] 7    70   700
[2,] 8    80   800
[3,] 9    90   900
```

Be careful if the list elements have names. In that case, do.call interprets the element names as names of parameters to the function, which might cause trouble.

This recipe presents the most basic use of do.call. The function is quite powerful and has many other uses. See the help page for more details.

See Also

See Recipe 5.33 for converting between datatypes.

12.19 Defining Your Own Binary Operators

Problem

You want to define your own binary operators, making your R code more streamlined and readable.

Solution

R recognizes any text between percent signs (%...%) as a binary operator. Create and define a new binary operator by assigning a two-argument function to it.

Discussion

R contains an interesting feature that lets you define your own binary operators. Any text between two percentage signs (%...%) is automatically interpreted by R as a binary operator. R predefines several such operators, such as %/% for integer division and %*% for matrix multiplication.

You can create a new binary operator by assigning a function to it. This example creates an operator, %+-%:

```
> '%+-%' <- function(x,margin) x + c(-1,+1)*margin
```

The expression x %+-% m calculates x ± m. Here it calculates $100 \pm (1.96 \times 15)$, the two-standard-deviation range of a standard IQ test:

```
> 100 %+-% (1.96*15)
[1]  70.6 129.4
```

Notice that we quote the binary operator when defining it but not when using it.

The pleasure of defining your own operators is that you can wrap commonly used operations inside a succinct syntax. If your application frequently concatenates strings without an intervening blank, then you might define a binary concatenation operator for that purpose:

```
> '%+%' <- function(s1,s2) paste(s1,s2,sep="")
> "Hello" %+% "World"
[1] "HelloWorld"
> "limit=" %+% round(qnorm(1-0.05/2), 2)
[1] "limit=1.96"
```

A danger of defining your own operators is that the code becomes less portable to other environments. Bring the definitions along with the code in which they are used; otherwise, R will complain about undefined operators.

All user-defined operators have the same precedence and are listed collectively in Table 2-1 as *%any%*. Their precedence is fairly high: higher than multiplication and division but lower than exponentiation and sequence creation. As a result, it's easy to misexpress yourself. If we omit parentheses from the "%+-%" example, we get an unexpected result:

```
> 100 %+-% 1.96*15
[1] 1470.6 1529.4
```

R interpreted the expression as (100 %+-% 1.96) * 15.

See Also

See Recipe 2.11 for more about operator precedence and Recipe 2.12 for how to define a function.

Beyond Basic Numerics and Statistics

Introduction

This chapter presents a few advanced techniques such as those you might encounter in the first or second year of a graduate program in applied statistics.

Most of these recipes use functions available in the base distribution. Through add-on packages, R provides some of the world's most advanced statistical techniques. This is because researchers in statistics now use R as their *lingua franca*, showcasing their newest work. Anyone looking for a cutting-edge statistical technique is urged to search CRAN and the Web for possible implementations.

13.1 Minimizing or Maximizing a Single-Parameter Function

Problem

Given a single-parameter function f, you want to find the point at which f reaches its minimum or maximum.

Solution

To minimize a single-parameter function, use `optimize`. Specify the function to be minimized and the bounds for its domain (x):

```
> optimize(f, lower=lowerBound, upper=upperBound)
```

If you instead want to maximize the function, specify `maximum=TRUE`:

```
> optimize(f, lower=lowerBound, upper=upperBound, maximum=TRUE)
```

Discussion

The `optimize` function can handle functions of one argument. It requires upper and lower bounds for x that delimit the region to be searched. The following example finds the minimum of a polynomial, $3x^4 - 2x^3 + 3x^2 - 4x + 5$:

```
> f <- function(x) 3*x^4 - 2*x^3 + 3*x^2 - 4*x + 5
> optimize(f, lower=-20, upper=20)
$minimum
[1] 0.5972778

$objective
[1] 3.636756
```

The returned value is a list with two elements: minimum, the x value that minimizes the function; and objective, the value of the function at that point.

A tighter range for lower and upper means a smaller region to be searched and hence a faster optimization. However, if you are unsure of the appropriate bounds then use big-but-reasonable values such as lower=-1000 and upper=1000. Just be careful that your function does not have multiple minima within that range! The optimize function will find and return only one such minimum.

See Also

See Recipe 13.2.

13.2 Minimizing or Maximizing a Multiparameter Function

Problem

Given a multiparameter function f, you want to find the point at which f reaches its minimum or maximum.

Solution

To minimize a multiparameter function, use optim. You must specify the starting point, which is a vector of initial arguments for f:

```
> optim(startingPoint, f)
```

To maximize the function instead, specify this control parameter:

```
> optim(startingPoint, f, control=list(fnscale=-1))
```

Discussion

The optim function is more general than optimize (Recipe 13.1) because it handles multiparameter functions. To evaluate your function at a point, optim packs the point's coordinates into a vector and calls your function with that vector. The function should return a scalar value. optim will begin at your starting point and move through the parameter space, searching for the function's minimum.

Here is an example of using `optim` to fit a nonlinear model. Suppose you believe that the paired observations z and x are related by $z_i = (x_i + \alpha)^\beta + \varepsilon_i$, where α and β are unknown parameters and where the ε_i are nonnormal noise terms. Let's fit the model by minimizing a robust metric, the sum of the absolute deviations:

$$\Sigma |z - (x + a)^b|$$

First we define the function to be minimized. Observe that the function has only one formal parameter, a two-element vector. The actual parameters to be evaluated, a and b, are packed into the vector in locations 1 and 2:

```
f <- function(v) { a <- v[1]; b <- v[2]        # "Unpack" v, giving a and b
                   sum(abs(z - ((x + a)^b)))
                 }
```

This call to `optim` starts from $(1, 1)$ and searches for the minimum point of f:

```
> optim(c(1,1), f)
$par
[1] 10.0072591  0.6997801

$value
[1] 1.26311

$counts
function gradient
     485       NA

$convergence
[1] 0

$message
NULL
```

The returned list includes `convergence`, the indicator of success or failure. If this indicator is 0 then `optim` found a minimum; otherwise, it did not. Obviously, the convergence indicator is the most important returned value because other values are meaningless if the algorithm did not converge.

The returned list also includes `par`, the parameters that minimize our function; and `value`, the value of f at that point. In this case, `optim` did converge and found a minimum point at approximately $a = 10.01$ and $b = 0.700$.

There are no lower and upper bounds for `optim`, just the starting point that you provide. A better guess for the starting point means a faster minimization.

The `optim` function supports several different minimization algorithms, and you can select among them. If the default algorithm does not work for you, see the help page for alternatives. A typical problem with multidimensional minimization is that the algorithm gets stuck at a local minimum and fails to find a deeper, global minimum. Generally speaking, the algorithms that are more powerful are less likely to get stuck. However, there is a trade-off: they also tend to run more slowly.

See Also

The R community has implemented many tools for optimization. On CRAN, see the task view for Optimization and Mathematical Programming (*http://cran.r-project.org/web/views/Optimization.html*) for more solutions.

13.3 Calculating Eigenvalues and Eigenvectors

Problem

You want to calculate the eigenvalues or eigenvectors of a matrix.

Solution

Use the eigen function. It returns a list with two elements, values and vectors, which contain (respectively) the eigenvalues and eigenvectors.

Discussion

Suppose we have a matrix such as the Fibonacci matrix:

```
> fibmat
     [,1] [,2]
[1,]    0    1
[2,]    1    1
```

Given the matrix, the eigen function will return a list of its eigenvalues and eigenvectors:

```
> eigen(fibmat)
$values
[1]  1.618034 -0.618034

$vectors
          [,1]       [,2]
[1,] 0.5257311 -0.8506508
[2,] 0.8506508  0.5257311
```

Use either eigen(fibmat)$values or eigen(fibmat)$vectors to select the needed value from the list.

13.4 Performing Principal Component Analysis

Problem

You want to identify the principal components of a multivariable dataset.

Solution

Use the prcomp function. The first argument is a formula whose righthand side is the set of variables, separated by plus signs (+). The lefthand side is empty:

```
> r <- prcomp( ~ x + y + z)
> summary(r)
```

Discussion

The base software for R includes two functions for principal components analysis, prcomp and princomp. The documentation mentions that prcomp has better numerical properties, so that's the function presented here.

An important use of principal components analysis (PCA) is to reduce the dimensionality of your dataset. Suppose your data contains a large number N of variables. Ideally, all the variables are more-or-less independent and contributing equally. But if you suspect that some variables are redundant, the PCA can tell you the number of sources of variance in your data. If that number is near N, then all the variables are useful. If the number is less than N, then your data can be reduced to a dataset of smaller dimensionality.

The PCA recasts your data into a vector space where the first dimension captures the most variance, the second dimension captures the second most, and so forth. The actual output from prcomp is an object that, when printed, gives the needed vector rotation:

```
> r <- prcomp( ~ x + y)
> print(r)
Standard deviations:
[1] 0.3724471 0.1306414

Rotation:
          PC1        PC2
x -0.5312726 -0.8472010
y -0.8472010  0.5312726
```

I find the summary of the PCA much more useful. It shows the proportion of variance that is captured by each component:

```
> summary(r)
Importance of components:
                          PC1   PC2
Standard deviation      0.372 0.131
Proportion of Variance  0.890 0.110
Cumulative Proportion   0.890 1.000
```

In this example, the first component captured 89% of the variance and the second component only 11%, so we know the first component captured most of it.

After calling prcomp, use plot(r) to view a bar chart of the variances of the principal components and predict(r) to rotate your data to the principal components.

See Also

See Recipe 13.9 for an example of using principal components analysis. Further uses of PCA in R are discussed in *Modern Applied Statistics with S-Plus* (Venables and Ripley, Springer).

13.5 Performing Simple Orthogonal Regression

Problem

You want to create a linear model using orthogonal regression in which the variances of *x* and *y* are treated symmetrically.

Solution

Use `prcomp` to perform principal components analysis on *x* and *y*. From the resulting rotation, compute the slope and intercept:

```
> r <- prcomp( ~ x + y )
> slope <- r$rotation[2,1] / r$rotation[1,1]
> intercept <- r$center[2] - slope*r$center[1]
```

Discussion

Orthogonal regression is also known as *total least squares* (TLS).

The ordinary least squares (OLS) algorithm has an odd property: it is asymmetric. That is, calculating `lm(y ~ x)` is not the mathematical inverse of calculating `lm(x ~ y)`. The reason is that OLS assumes the *x* values to be constants and the *y* values to be random variables, so all the variance is attributed to *y* and none is attributed to *x*. This creates an asymmetric situation.

The asymmetry is illustrated in Figure 13-1, whose upper-left panel illustrates the fit for `lm(y ~ x)`. The OLS algorithm tries the minimize the vertical distances, which are shown as dotted lines. The upper-right panel shows the identical dataset but fit with `lm(x ~ y)` instead, so the algorithm is minimizing the horizontal dotted lines. Obviously, you'll get a different result depending upon which distances are minimized.

The lower panel of Figure 13-1 is quite different. It uses PCA to implement orthogonal regression. Now the distances being minimized are the orthogonal distances from the data points to the regression line. This is a symmetric situation: reversing the roles of *x* and *y* does not change the distances to be minimized.

Implementing a basic orthogonal regression in R is quite simple. First, perform the PCA:

```
> r <- prcomp( ~ x + y)
```

Next, use the rotations to compute the slope:

```
> slope <- r$rotation[2,1] / r$rotation[1,1]
```

and then, from the slope, calculate the intercept:

```
> intercept <- r$center[2] - slope*r$center[1]
```

I call this a "basic" regression because it yields only the point estimates for the slope and intercept, not the confidence intervals. Obviously, we'd like to have the regression statistics, too. Recipe 13.8 shows one way to estimate the confidence intervals using a bootstrap algorithm.

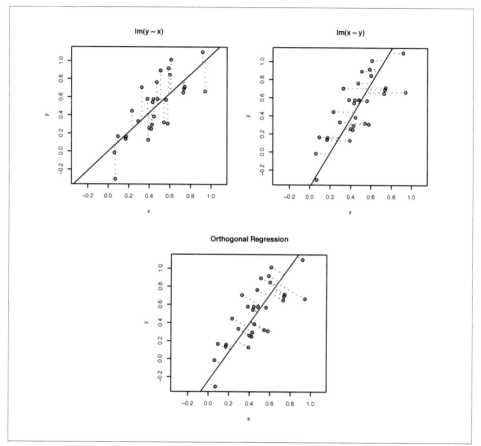

Figure 13-1. Ordinary least squares versus orthogonal regression

See Also

Principal components analysis is described in Recipe 13.4. The graphics in Figure 13-1 were inspired by the work of Vincent Zoonekynd and his tutorial on regression at *http://zoonek2.free.fr/UNIX/48_R/09.html*.

13.6 Finding Clusters in Your Data

Problem

You believe your data contains *clusters*: groups of points that are "near" each other. You want to identify those clusters.

Solution

Your dataset, *x*, can be a vector, data frame, or matrix. Assume that *n* is the number of clusters you desire:

```
> d <- dist(x)              # Compute distances between observations
> hc <- hclust(d)           # Form hierarchical clusters
> clust <- cutree(hc, k=n)  # Organize them into the n largest clusters
```

The result, clust, is a vector of numbers between 1 and *n*, one for each observation in *x*. Each number classifies its corresponding observation into one of the *n* clusters.

Discussion

The dist function computes distances between all the observations. The default is Euclidean distance, which works well for many applications, but other distance measures are also available.

The hclust function uses those distances to form the observations into a hierarchical tree of clusters. You can plot the result of hclust to create a visualization of the hierarchy, called a *dendrogram*, as shown in Figure 13-2.

Finally, cutree extracts clusters from that tree. You must specify either how many clusters you want or the height at which the tree should be cut. Often the number of clusters is unknown, in which case you will need to explore the dataset for clustering that makes sense in your application.

I'll illustrate clustering of a synthetic dataset. We start by generating 99 normal variates, each with a randomly selected mean of either –3, 0, or +3:

```
> means <- sample(c(-3,0,+3), 99, replace=TRUE)
> x <- rnorm(99, mean=means)
```

For our own curiosity, we can compute the true means of the original clusters. (In a real situation, we would not have the means factor and would be unable to perform this computation.) We can confirm that the groups' means are pretty close to –3, 0, and +3:

```
> tapply(x, factor(means), mean)
        -3          0          3
-3.0151424 -0.2238356  2.7602507
```

To discover the clusters, we first compute the distances between all points:

```
> d <- dist(x)
```

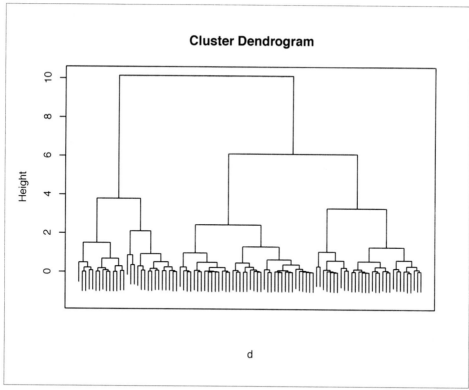

Figure 13-2. Plotting the cluster hierarchy

Then we create the hierarchical clusters:

```
> hc <- hclust(d)
```

Now we can extract the three largest clusters. Obviously, we have a huge advantage here because we know the true number of clusters. Real life is rarely that easy:

```
> clust <- cutree(hc, k=3)
```

clust is a vector of integers between 1 and 3, one integer for each observation in the sample, that assigns each observation to a cluster. Here are the first 20 cluster assignments:

```
> head(clust,20)
 [1] 1 2 2 2 1 2 3 3 2 3 1 3 2 3 2 1 2 1 1 3
```

By treating the cluster number as a factor, we can compute the mean of each statistical cluster (Recipe 6.5):

```
> tapply(x, clust, mean)
         1          2          3
 3.1899159 -2.6992682  0.2363639
```

R did a good job of splitting data into clusters: the means appear distinct, with one near –2.7, one near 0.27, and one near +3.2. (The order of the extracted means does not necessarily match the order of the original groups, of course.) The extracted means are similar but not identical to the original means. Side-by-side box plots can show why:

```
> plot(x ~ factor(means), main="Original Clusters", xlab="Cluster Mean")
> plot(x ~ factor(clust), main="Identified Clusters", xlab="Cluster Number")
```

The box plots are shown in Figure 13-3. The clustering algorithm perfectly separated the data into nonoverlapping groups. The original clusters overlapped, whereas the identified clusters do not.

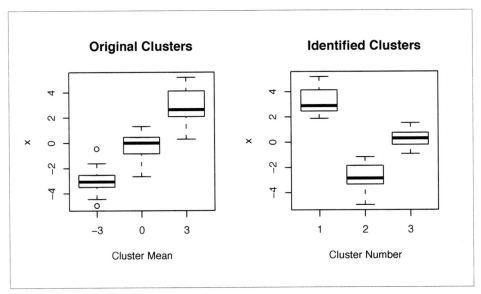

Figure 13-3. Clusters comparison

This illustration used one-dimensional data, but the `dist` function works equally well on multidimensional data stored in a data frame or matrix. Each row in the data frame or matrix is treated as one observation in a multidimensional space, and `dist` computes the distances between those observations.

See Also

This demonstration is based on the clustering features of the base package. There are other packages such as `mclust` that offer alternative clustering mechanisms.

13.7 Predicting a Binary-Valued Variable (Logistic Regression)

Problem

You want to perform *logistic regression*, a regression model that predicts the probability of a binary event occurring.

Solution

Call the `glm` function with `family=binomial` to perform logistic regression. The result is a model object:

```
> m <- glm(b ~ x1 + x2 + x3, family=binomial)
```

Here, `b` is a factor with two levels (e.g., `TRUE` and `FALSE`, 0 and 1) `x1`, `x2`, and `x3` are predictor variables.

Use the model object, `m`, and the `predict` function to predict a probability from new data:

```
> dfrm <- data.frame(x1=value, x2=value, x3=value)
> predict(m, type="response", newdata=dfrm)
```

Discussion

Predicting a binary-valued outcome is a common problem in modeling. Will a treatment be effective or not? Will prices rise or fall? Who will win the game, team A or team B? Logistic regression is useful for modeling these situations. In the true spirit of statistics, it does not simply give a "thumbs up" or "thumbs down" answer; rather, it computes a probability for each of the two possible outcomes.

In the call to `predict`, we set `type="response"` so that `predict` returns a probability. Otherwise, it returns log-odds, which I don't find as useful.

In his unpublished book entitled *Practical Regression and ANOVA Using R*, Faraway gives an example of predicting a binary-valued variable: `test` from the dataset `pima` is true if the patient tested positive for diabetes. The predictors are diastolic blood pressure and body mass index (BMI). Faraway uses linear regression, so let's try logistic regression instead:

```
> data(pima, package="faraway")
> b <- factor(pima$test)
> m <- glm(b ~ diastolic + bmi, family=binomial, data=pima)
```

The summary of the resulting model, `m`, shows that the respective *p*-values for the `diastolic` and the `bmi` variables are 0.805 and (essentially) zero. We can therefore conclude that only the `bmi` variable is significant:

```
> summary(m)

Call:
glm(formula = b ~ diastolic + bmi, family = binomial, data = pima)
```

```
Deviance Residuals:
    Min      1Q   Median      3Q     Max
-1.9128  -0.9180  -0.6848  1.2336  2.7417

Coefficients:
             Estimate Std. Error z value Pr(>|z|)
(Intercept) -3.629553   0.468176  -7.753 9.01e-15 ***
diastolic   -0.001096   0.004432  -0.247    0.805
bmi          0.094130   0.012298   7.654 1.95e-14 ***
---
Signif. codes:  0 '***' 0.001 '**' 0.01 '*' 0.05 '.' 0.1 ' ' 1

(Dispersion parameter for binomial family taken to be 1)

    Null deviance: 993.48  on 767  degrees of freedom
Residual deviance: 920.65  on 765  degrees of freedom
AIC: 926.65

Number of Fisher Scoring iterations: 4
```

Because only the bmi variable is significant, we can create a reduced model like this:

```
> m.red <- glm(b ~ bmi, family=binomial, data=pima)
```

Let's use the model to calculate the probability that someone with an average BMI (32.0) will test positive for diabetes:

```
> newdata <- data.frame(bmi=32.0)
> predict(m.red, type="response", newdata=newdata)
        1
0.3332689
```

According to this model, the probability is about 33.3%. The same calculation for someone in the 90th percentile gives a probability of 54.9%:

```
> newdata <- data.frame(bmi=quantile(pima$bmi, .90))
> predict(m.red, type="response", newdata=newdata)
      90%
0.5486215
```

See Also

Using logistic regression involves interpreting the deviance to judge the significance of the model. I suggest reviewing a text on logistic regression before attempting to draw any conclusions from your regression.

13.8 Bootstrapping a Statistic

Problem

You have a dataset and a function to calculate a statistic from that dataset. You want to estimate a confidence interval for the statistic.

Solution

Use the boot package. Apply the boot function to calculate bootstrap replicates of the statistic:

```
> library(boot)
> bootfun <- function(data, indices) {
+     # . . . calculate statistic using data[indices] . . .
+     return(statistic)
+ }

> reps <- boot(data, bootfun, R=999)
```

Here, *data* is your original data set, which can be stored in either a vector or a data frame. The statistic function (bootfun in this case) should expect two arguments: data is the original dataset, and indices is a vector of integers that selects the bootstrap sample from data.

Next, use the boot.ci function to estimate a confidence interval from the replications:

```
> boot.ci(reps, type=c("perc","bca"))
```

Discussion

Anybody can calculate a statistic, but that's just the point estimate. We want to take it to the next level: What is the confidence interval? For some statistics, we can calculate the confidence interval (CI) analytically. The confidence interval for a mean, for instance, is calculated by the t.test function. Unfortunately, that is the exception and not the rule. For most statistics, the mathematics is too tortuous or simply unknown, and there is no known closed-form calculation for the confidence interval.

The bootstrap algorithm can estimate a confidence interval even when no closed-form calculation is available. It works like this. The algorithm assumes that you have a sample of size N and a function to calculate the statistic:

1. Randomly select N elements from the sample, *sampling with replacement*. That set of elements is called a *bootstrap sample*.

2. Apply the function to the bootstrap sample to calculate the statistic. That value is called a *bootstrap replication*.

3. Repeat steps 1 and 2 many times to yield many (typically thousands) of bootstrap replications.

4. From the bootstrap replications, compute the confidence interval.

That last step may seem mysterious, but there are several algorithms for computing the confidence interval. A simple one uses percentiles of the replications, such as taking the 2.5 percentile and the 97.5 percentile to form the 95% confidence interval.

I am a huge fan of the bootstrap because I work daily with obscure statistics, it is important that I know their confidence interval, and there is definitely no known formula for their CI. The bootstrap gives me a good approximation.

Let's work an example. In Recipe 13.4 we estimated the slope of a line using orthogonal regression. That gave us a point estimate, but how can we find the confidence interval? First, we encapsulate the slope calculation within a function:

```
> stat <- function(data, indices) {
+               r <- prcomp( ~ x + y, data=data, subset=indices)
+               slope <- r$rotation[2,1] / r$rotation[1,1]
+               return(slope)
+ }
```

Notice that the function is careful to select the subset defined by indices and to compute the slope from that exact subset.

Next, we calculate 999 replications of the slope. Recall that we had two vectors, x and y, in the original recipe; here, we combine them into a data frame:

```
> boot.data <- data.frame(x=x, y=y)
> reps <- boot(boot.data, stat, R=999)
```

The choice of 999 replications is a good starting point. You can always repeat the bootstrap with more and see if the results change significantly.

The boot.ci function can estimate the CI from the replications. It implements several different algorithms, and the type argument selects which algorithms are performed. For each selected algorithm, boot.ci will return the resulting estimate:

```
> boot.ci(reps, type=c("perc","bca"))
BOOTSTRAP CONFIDENCE INTERVAL CALCULATIONS
Based on 999 bootstrap replicates

CALL :
boot.ci(boot.out = reps, type = c("perc", "bca"))

Intervals :
Level      Percentile              BCa
95%   ( 1.144,  2.277 )    ( 1.123,  2.270 )
Calculations and Intervals on Original Scale
```

Here I chose two algorithms, percentile and BCa, by setting type=c("perc","bca"). The two resulting estimates appear at the bottom under their names. Other algorithms are available; see the help page.

You will note that the two confidence intervals are slightly different: (1.144, 2.277) versus (1.123, 2.270). This is an uncomfortable but inevitable result of using two different algorithms. I don't know any method for deciding which is better. If the selection is a critical issue, you will need to study the reference and understand the differences. In the meantime, my best advice is to be conservative and use the wider interval; in this case, that would be (1.123, 2.270).

By default, boot.ci estimates a 95% confidence interval. You can change that via the conf argument, like this:

```
> boot.ci(reps, type=c("perc","bca"), conf=0.90)
```

See Also

See Recipe 13.4 for the slope calculation. A good tutorial and reference for the bootstrap algorithm is *An Introduction to the Bootstrap* by Efron and Tibshirani (Chapman & Hall/CRC).

13.9 Factor Analysis

Problem

You want to perform factor analysis on your dataset, usually to discover what your variables have in common.

Solution

Use the `factanal` function, which requires your dataset and your estimate of the number of factors:

```
> factanal(data, factors=n)
```

The output includes n factors, showing the loadings of each input variable for each factor.

The output also includes a *p*-value. Conventionally, a *p*-value of less than 0.05 indicates that the number of factors is too small and does not capture the full dimensionality of the dataset; a *p*-value exceeding 0.05 indicates that there are likely enough (or more than enough) factors.

Discussion

Factor analysis creates linear combinations of your variables, called *factors*, that abstract the variable's underlying commonality. If your N variables are perfectly independent, then they have nothing in common and N factors are required to describe them. But to the extent that the variables have an underlying commonality, fewer factors capture most of the variance and so fewer than N factors are required.

For each factor and variable, we calculate the correlation between them and call it the *loading*. Variables with a high loading are well explained by the factor. We can square the loading to know what fraction of the variable's total variance is explained by the factor.

Factor analysis is useful when it shows that a few factors capture most of the variance of your variables. Thus, it alerts you to redundancy in your data. In that case you can reduce your dataset by combining closely related variables or by eliminating redundant variables altogether.

A more subtle application of factor analysis is interpreting the factors to find interrelationships between your variables. If two variables both have large loadings for the same factor, then you know they have something in common. What is it? There is no mechanical answer. You'll need to study the data and its meaning.

There are two tricky aspects of factor analysis. The first is choosing the number of factors. Fortunately, you can use principal components analysis to get a good initial estimate of the number of factors. The second tricky aspect is interpreting the factors themselves.

I'll illustrate factor analysis by using stock prices—or, more precisely, changes in stock prices. The dataset contains 6 months of price changes for the stocks of 12 companies. Every company is involved in the petroleum and gasoline industry. Their stock prices probably move together, since they are subject to the similar economic and market forces. We might ask: How many factors are required to explain their changes? If only one factor is required, then all the stocks are the same and one is as good as another. If many factors are required, we know that owning several of them provides diversification.

We start by doing a principal components analysis on `diffs`, the data frame of price changes. Plotting the PCA results shows the variance captured by the components:

```
> plot(prcomp(diffs))
```

The plot is shown in Figure 13-4. We can see that the first two components capture much of the variance, but possibly a third is required. So we perform the initial factor analysis while assuming that two factors are required:

```
> factanal(diffs, factors=2)

Call:
factanal(x = diffs, factors = 2)

.
.  (etc.)
.

Test of the hypothesis that 2 factors are sufficient.
The chi square statistic is 102.25 on 43 degrees of freedom.
The p-value is 9.83e-07
```

We can ignore most of the output because the p-value at the bottom is essentially zero (9.83×10^{-7}). The small p-value indicates that two factors are insufficient, so the analysis isn't good. More are required, so we try again with three factors instead:

```
> factanal(diffs, factors=3)

Call:
factanal(x = diffs, factors = 3)

Uniquenesses
  APC    BP   BRY   CVX   HES   MRO   NBL   OXY   SUN   TOT   VLO   XOM
```

0.005 0.509 0.299 0.162 0.150 0.207 0.321 0.265 0.098 0.185 0.257 0.155

```
Loadings:
    Factor1 Factor2 Factor3
APC 0.286   0.176   0.939
BP  0.212   0.150   0.651
BRY 0.628   0.474   0.286
CVX 0.819   0.320   0.253
HES 0.751   0.415   0.338
MRO 0.721   0.396   0.342
NBL 0.606   0.382   0.406
OXY 0.713   0.436   0.192
SUN 0.394   0.847   0.172
TOT 0.786   0.372   0.241
VLO 0.465   0.672   0.275
XOM 0.833   0.218   0.322

               Factor1 Factor2 Factor3
SS loadings      4.835   2.400   2.151
Proportion Var   0.403   0.200   0.179
Cumulative Var   0.403   0.603   0.782
```

Test of the hypothesis that 3 factors are sufficient.
The chi square statistic is 35.74 on 33 degrees of freedom.
The p-value is 0.341

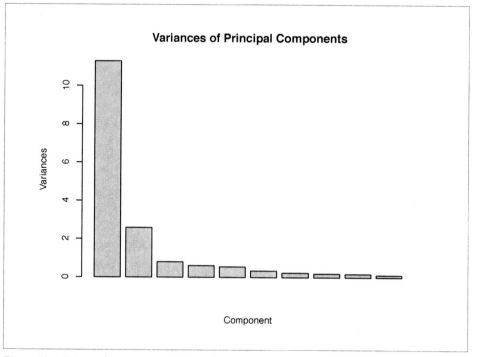

Figure 13-4. Principal components of price changes

The large *p*-value (0.341) confirms that three factors are sufficient, so we can use the analysis.

The output includes a table of explained variance, shown here:

	Factor1	Factor2	Factor3
SS loadings	4.835	2.400	2.151
Proportion Var	0.403	0.200	0.179
Cumulative Var	0.403	0.603	0.782

This table shows that the proportion of variance explained by each factor is 0.403, 0.200, and 0.179, respectively. Cumulatively, they explain 0.782 of the variance, which leaves $1 - 0.782 = 0.218$ unexplained.

Next we want to interpret the factors, which is more like voodoo than science. Let's look at the loadings, repeated here:

```
Loadings:
    Factor1 Factor2 Factor3
APC 0.286   0.176   0.939
BP  0.212   0.150   0.651
BRY 0.628   0.474   0.286
CVX 0.819   0.320   0.253
HES 0.751   0.415   0.338
MRO 0.721   0.396   0.342
NBL 0.606   0.382   0.406
OXY 0.713   0.436   0.192
SUN 0.394   0.847   0.172
TOT 0.786   0.372   0.241
VLO 0.465   0.672   0.275
XOM 0.833   0.218   0.322
```

Each row is labeled with the variable name (stock symbol): APC, BP, BRY, and so forth. The first factor has many large loadings, indicating that it explains the variance of many stocks. This is a common phenomenon in factor analysis. We are often looking at related variables, and the first factor captures their most basic relationship. In this example, we are dealing with stocks, and most stocks move together in concert with the broad market. That's probably captured by the first factor.

The second factor is more subtle. Notice that the loadings for SUN (0.847) and VLO (0.672) are the dominant ones and that all other stocks have noticeably smaller loadings. This indicates a connection between SUN and VLO. Perhaps they operate together in a common area (e.g., retail sales) and so tend to move together.

The third factor also has two dominant loadings: APC (0.939) and BP (0.651). During the sample period, there was a horrific environmental disaster—caused by an oil rig explosion in the Gulf of Mexico—that cast a dark shadow on the future of oil exploration in the Gulf. British Petroleum (BP) owned the rig, and Anadarko Petroleum (APC) is a major supplier of equipment to Gulf operators. Those two stocks moved in unison while the market assessed the disaster's impact on BP and APC.

In summary, it seems there are three groups of stocks here:

- SUN and VLO
- APC and BP
- Everything else

Factor analysis is an art and a science. I suggest that you read a good book on multivariate analysis before employing it.

See Also

See Recipe 13.4 for more about principal components analysis.

Time Series Analysis

Introduction

Time series analysis has become a hot topic with the rise of quantitative finance and automated trading of securities. Many of the facilities described in this chapter were invented by practitioners and researchers in finance, securities trading, and portfolio management.

Before starting any time series analysis in R, a key decision is your choice of data representation (object class). This is especially critical in an object-oriented language such as R because the choice affects more than how the data is stored; it also dictates which functions (methods) will be available for loading, processing, analyzing, printing, and plotting your data. When I first started using R, for example, I simply stored my time series data in vectors. That seemed natural. However, I quickly discovered that none of the coolest analytics for time series analysis worked with simple vectors. I switched to using an object class intended for time series analysis, and this was the gateway to important functions and analytics.

This chapter's first recipe recommends using the zoo or xts packages for representing time series data. They are quite general and should meet the needs of most users. Nearly every subsequent recipe assumes you are using one of those two representations.

 The xts implementation is a superset of zoo, so xts can do everything that zoo can do. In this chapter, whenever a recipe works for a zoo object, you can safely assume (unless stated otherwise) that it also works for an xts object.

The base distribution of R includes a time series class called ts. I don't recommend this representation for general use because the implementation itself is too restrictive and because the associated functions (methods) are limited in scope, usefulness, and power.

Date Versus Datetime

Every observation in a time series has an associated date or time. The object classes used in this chapter, zoo and xts, give you the choice of using either dates or datetimes for representing the data's time component. You would use dates to represent daily data, of course, and also for weekly, monthly, or even annual data; in these cases, the date gives the day on which the observation occurred. You would use datetimes for intraday data, where both the date and time of observation are needed.

In describing this chapter's recipes, it's pretty cumbersome to keep saying "date or datetime." So I simplified the prose by assuming that your data are daily and thus use whole dates. Please bear in mind, of course, that you are free and able to use timestamps below the resolution of a calendar date.

See Also

R has many useful functions and packages for time series analysis. You'll find pointers to them in the task view for Time Series Analysis (*http://cran.r-project.org/web/views/TimeSeries.html*).

14.1 Representing Time Series Data

Problem

You want an R data structure that can represent time series data.

Solution

I recommend the zoo and xts packages. They define a data structure for time series, and they contain many useful functions for working with time series data. Create a zoo object this way, where x is a vector or data frame and dt is a vector of corresponding dates or datetimes:

```
> library(zoo)
> ts <- zoo(x, dt)
```

Create an xts object in this way:

```
> library(xts)
> ts <- xts(x, dt)
```

Convert between representations of the time series data by using as.zoo and as.xts:

as.zoo(ts)
> Converts ts to a zoo object

as.xts(ts)
> Converts ts to an xts object

Discussion

R has at least eight different implementations of data structures for representing time series. I haven't tried them all, but I can say that the zoo and xts packages are excellent packages for working with time series data and are better than the others that I have tried.

These representations assume you have two vectors: a vector of observations (data) and a vector of dates or times of those observations. The zoo function combines them into a zoo object:

```
> ts <- zoo(x, dt)
```

The xts function is similar, returning an xts object:

```
> ts <- xts(x, dt)
```

The data, x, should be numeric. The vector of dates or datetimes, dt, is called the *index*. Legal indices vary between the packages:

zoo

> The index can be any ordered values, such as Date objects, POSIXct objects, integers, or even floating-point values.

xts

> The index must be a supported date or time class. This includes Date, POSIXct, and chron objects. Those should be sufficient for most applications, but you can also use yearmon, yearqtr, and dateTime objects. The xts package is more restrictive than zoo because it implements powerful operations that require a time-based index.

The following example creates a zoo object that contains the price of IBM stock for the first five days of 2010; it uses Date objects for the index:

```
> prices <- c(132.45, 130.85, 130.00, 129.55, 130.85)
> dates <- as.Date(c("2010-01-04", "2010-01-05", "2010-01-06",
+                     "2010-01-07","2010-01-08"))
> ibm.daily <- zoo(prices, dates)
> print(ibm.daily)
2010-01-04 2010-01-05 2010-01-06 2010-01-07 2010-01-08
    132.45     130.85     130.00     129.55     130.85
```

In contrast, the next example captures the price of IBM stock at one-second intervals. It represents time by the number of hours past midnight starting at 9:30 a.m. (1 second ≈ 0.00027778 hours):

```
> prices <- c(131.18, 131.20, 131.17, 131.15, 131.17)
> seconds <- c(9.5, 9.500278, 9.500556, 9.500833, 9.501111)
> ibm.sec <- zoo(prices, seconds)
> print(ibm.sec)
   9.5 9.5003 9.5006 9.5008 9.5011
131.18 131.20 131.17 131.15 131.17
```

Those two examples used a single time series, where the data came from a vector. Both zoo and xts can also handle multiple, parallel time series. For this, capture the several time series in a data frame and then create a multivariate time series by calling the zoo (or xts) function:

```
> ts <- zoo(dfrm, dt)      # OR: ts <- xts(dfrm, dt)
```

The second argument is a vector of dates (or datetimes) for each observation. There is only one vector of dates for all the time series; in other words, all observations in each row of the data frame must have the same date. See Recipe 14.5 if your data have mismatched dates.

Once the data is captured inside a zoo or xts object, you can extract the pure data via coredata, which returns a simple vector (or matrix):

```
> coredata(ibm.daily)
[1] 132.45 130.85 130.00   0.00 129.55
```

You can extract the date or time portion via index:

```
> index(ibm.daily)
[1] "2010-01-04" "2010-01-05" "2010-01-06" "2010-01-07" "2010-01-08"
```

The xts package is strongly similar to zoo. It is optimized for speed and so is especially well suited for processing large volumes of data. It is also clever about converting to and from other time series representations.

One big advantage of capturing data inside a zoo or xts object is that special-purpose functions become available for printing, plotting, differencing, merging, periodic sampling, applying rolling functions, and other useful operations. There is even a function, read.zoo, dedicated to reading time series data from ASCII files.

Remember that the xts package can do everything that the zoo package can do, so everywhere that this chapter talks about zoo objects you can also use xts objects.

If you are a serious user of time series data, I strongly recommend studying the documentation of these packages in order to learn about the ways they can improve your life. They are rich packages with many useful features.

See Also

See CRAN for documentation on zoo (*http://cran.r-project.org/web/packages/zoo/*) and xts (*http://cran.r-project.org/web/packages/xts/*); this documentation includes reference manuals, vignettes, and quick reference cards. If the packages are already installed on your computer, view their documentation using the vignette function:

```
> vignette("zoo")
> vignette("xts")
```

The timeSeries package is another good implementation of a time series object. It is part of the Rmetrics project for quantitative finance.

14.2 Plotting Time Series Data

Problem

You want to plot one or more time series.

Solution

Use plot(x), which works for zoo objects and xts objects containing either single or multiple time series.

For a simple vector v of time series observations, you can use either plot(v,type="l") or plot.ts(v).

Discussion

The generic plot function has a version for zoo objects and xts objects. It can plot objects that contain a single time series or multiple time series. In the latter case, it can plot each series in a separate plot or together in one plot.

Suppose that ibm.infl is a zoo object that contains two time series. One shows the quoted price of IBM stock during the 1970s, and the other is that same price adjusted for inflation.[*] If you plot the object with screens=1, R will plot the two time series together in one plot:

```
> plot(ibm.infl, screens=1)
```

If you specify screens=c(1,2), however, then R will plot both time series separately in two plots:

```
> plot(ibm.infl, screens=c(1,2))
```

The plot function provides a default label for the x-axis and y-axis, but they are not very informative. It does not provide a default title. So I almost always supply my own xlab, ylab, and main (title) parameters. This R code produces the plots shown in Figures 14-1 and 14-2:

```
xlab="Date"
ylab="Relative Price"
main="IBM: Historical vs. Inflation-Adjusted"

lty=c("dotted", "solid")
ylim=range(coredata(ibm.infl))

# Plot the two time series in two plots
plot(ibm.infl, screens=c(1,2), lty=lty, main=main, xlab=xlab, ylab=ylab, ylim=ylim)

# Plot the two time series in one plot
```

[*] The data shown do not match the historical stock quotes because they have been adjusted for dividends and stock splits, then normalized to a starting value of 100.

```
plot(ibm.infl, screens=1, lty=lty, main=main, xlab=xlab, ylab=ylab)

# Add a legend
legend(as.Date("1970-01-01"), 140,
       c("Hist","Infl-Adj"),
       lty=c("dotted", "solid"))
```

The code specifies two line types (lty) so that the two lines are drawn in two different styles, making them easier to distinguish. It also specifies the same y limits (ylim) for the separate plots, making them visually comparable.

I included both plots so you can experience their difference yourself. For my money, the two-in-one plot illustrates far more dramatically that inflation in the 1970s took a big bite out of IBM's stock price: when adjusted for inflation, that price actually fell.

See Also

For working with financial data, the quantmod package contains special plotting functions that produce beautiful, stylized plots.

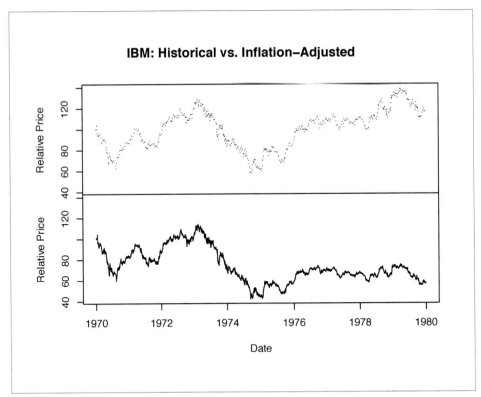

Figure 14-1. Plotting two time series with screen=c(1,2)

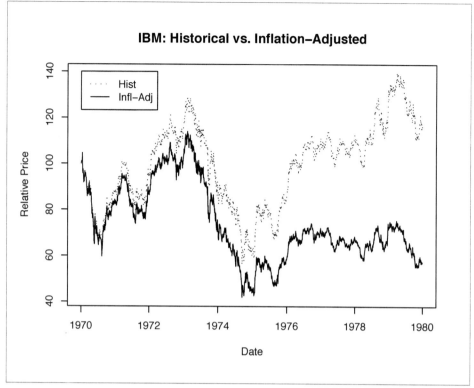

Figure 14-2. Plotting two time series with screen=1

14.3 Extracting the Oldest or Newest Observations

Problem

You want to see only the oldest or newest observations of your time series.

Solution

Use head to view the oldest observations:

```
> head(ts)
```

Use tail to view the newest observations:

```
> tail(ts)
```

Discussion

The head and tail functions are generic, so they will work whether your data is stored in a simple vector, a zoo object, or an xts object.

Suppose you have a zoo object with a multiyear history of the price of IBM stock. You can't display the whole dataset because it would scroll off your screen. But you can view the initial observations:

```
> head(ibm)
1970-01-02 1970-01-05 1970-01-06 1970-01-07 1970-01-08 1970-01-09
    364.75     368.25     368.50     368.75     369.50     369.00
```

And you can view the final observations:

```
> tail(ibm)
1979-12-21 1979-12-24 1979-12-26 1979-12-27 1979-12-28 1979-12-31
     64.00      64.87      64.87      64.75      64.25      64.37
```

By default, head and tail show (respectively) the six oldest and six newest observations. You can see more observations by providing a second argument—for example, tail(ibm, 20).

The xts package also includes first and last functions, which use calendar periods instead of number of observations. We can use first and last to select data by number of days, weeks, months, or even years:

```
> first(as.xts(ibm), "3 weeks")
              [,1]
1970-01-02 364.75
1970-01-05 368.25
1970-01-06 368.50
1970-01-07 368.75
1970-01-08 369.50
1970-01-09 369.00
1970-01-12 367.75
1970-01-13 374.13
1970-01-14 373.75
1970-01-15 381.50
1970-01-16 369.75
> last(as.xts(ibm), "month")
              [,1]
1979-12-03 65.00
1979-12-04 65.37
1979-12-05 65.75
.
. (etc.)
.
1979-12-27 64.75
1979-12-28 64.25
1979-12-31 64.37
```

Notice that I converted the zoo object to an xts object by calling as.xts(ibm), forcing R to use the xts functions (methods). Also notice that a "week" is defined by the calendar, not just by any seven consecutive days.

See Also

See `help(first.xts)` and `help(last.xts)` for details of the `first` and `last` functions, respectively.

14.4 Subsetting a Time Series

Problem

You want to select one or more elements from a time series.

Solution

You can index a `zoo` or `xts` object by position. Use one or two subscripts, depending upon whether the object contains one time series or multiple time series:

`ts[i]`
> Selects the *i*th observation from a single time series

`ts[j,i]`
> Selects the *i*th observation of the *j*th time series of multiple time series

You can index the time series by a date object. Use the same type of object as the index of your time series. This example assumes that the index contains `Date` objects:

```
> ts[as.Date("yyyy-mm-dd")]
```

You can index it by a sequence of dates:

```
> dates <- seq(startdate, enddate, increment)
> ts[dates]
```

The `window` function can select a range by start and end date:

```
> window(ts, start=startdate, end=enddate)
```

Discussion

Recall our small sample of IBM stock prices:

```
> ibm.daily
2010-01-04 2010-01-05 2010-01-06 2010-01-07 2010-01-08
    132.45     130.85     130.00     129.55     130.85
```

We can select an observation by position, just like selecting elements from a vector (Recipe 2.9):

```
> ibm.daily[2]
2010-01-05
    130.85
```

We can also select multiple observations by position:

```
> ibm.daily[2:4]
2010-01-05 2010-01-06 2010-01-07
    130.85     130.00     129.55
```

Sometimes it's more useful to select by date. In this case, our index is built from Date objects, so we subscript the time series using a Date object (if the index were built from POSIXct objects, we'd use a POSIXct object):

```
> ibm.daily[as.Date('2010-01-05')]
2010-01-05
    130.85
```

We can select by a vector of Date objects:

```
> dates <- seq(as.Date('2010-01-04'), as.Date('2010-01-08'), by=2)
> ibm.daily[dates]
2010-01-04 2010-01-06 2010-01-08
    132.45     130.00     130.85
```

The window function is easier for selecting a range of consecutive dates:

```
> window(ibm.daily, start=as.Date('2010-01-05'), end=as.Date('2010-01-07'))
2010-01-05 2010-01-06 2010-01-07
    130.85     130.00     129.55
```

See Also

The xts package provides many other clever ways to index a time series. See the package documentation.

14.5 Merging Several Time Series

Problem

You have two or more time series. You want to merge them into a single time series object.

Solution

Use the zoo object to represent the time series; then use the merge function to combine them:

```
> merge(ts1, ts2)
```

Discussion

Merging two time series is an incredible headache when the two series have differing timestamps. Consider these two time series: the daily price of IBM stock from 1970 through 1979; and the monthly Consumer Price Index (CPI) for the same period:

```
> ibm
1970-01-02 1970-01-05 1970-01-06 1970-01-07 1970-01-08 1970-01-09 1970-01-12
    364.75      368.25      368.50      368.75      369.50      369.00      367.75
.
. (etc.)
.
1979-12-21 1979-12-24 1979-12-26 1979-12-27 1979-12-28 1979-12-31
     64.00       64.87       64.87       64.75       64.25       64.37
> cpi
1970-01-01 1970-02-01 1970-03-01 1970-04-01 1970-05-01 1970-06-01 1970-07-01
      37.8        38.0        38.2        38.5        38.6        38.8        39.0
.
. (etc.)
.
1979-05-01 1979-06-01 1979-07-01 1979-08-01 1979-09-01 1979-10-01 1979-11-01
      71.5        72.3        73.1        73.8        74.6        75.2        75.9
1979-12-01
      76.7
```

Obviously, the two time series have different timestamps because one is daily data and the other is monthly data. Even worse, the downloaded CPI data is timestamped for the first day of every month, even when that day is a holiday or weekend (e.g., New Year's Day).

Thank goodness for the merge function, which handles the messy details of reconciling the different dates:

```
> merge(ibm,cpi)
              ibm  cpi
1970-01-01    NA 37.8
1970-01-02 364.75   NA
1970-01-05 368.25   NA
1970-01-06 368.50   NA
1970-01-07 368.75   NA
1970-01-08 369.50   NA
.
. (etc.)
.
```

By default, merge finds the *union* of all dates: the output contains all dates from both inputs, and missing observations are filled with NA values. You can replace those NA values with the most recent observation by using the na.locf function from the zoo package:

```
> na.locf(merge(ibm,cpi)))
              ibm  cpi
1970-01-02 364.75 37.8
1970-01-05 368.25 37.8
1970-01-06 368.50 37.8
1970-01-07 368.75 37.8
1970-01-08 369.50 37.8
1970-01-09 369.00 37.8
.
. (etc.)
.
```

(Here locf stands for "last observation carried forward".) Observe that the NAs were replaced. Also, na.locf eliminated the first observation (1970-01-01) because there was no IBM stock price on that day.

You can get the *intersection* of all dates by setting all=FALSE:

```
> merge(ibm,cpi,all=FALSE)
              ibm  cpi
1970-04-01 337.75 38.5
1970-05-01 296.75 38.6
1970-06-01 287.00 38.8
1970-07-01 254.00 39.0
1970-09-01 263.88 39.2
1970-10-01 297.00 39.4
.
. (etc.)
.
```

Now the output is limited to observations that are *common* to both files.

Notice, however, that the intersection begins on April 1, not January 1. The reason is that January 1, February 1, and March 1 are all holidays or weekends, so there is no IBM stock price for those dates and hence no intersection with the CPI data. To fix this, see Recipe 14.6.

14.6 Filling or Padding a Time Series

Problem

Your time series data is missing observations. You want to fill or pad the data with the missing dates/times.

Solution

Create a zero-width (dataless) zoo object with the missing dates/times. Then merge your data with the zero-width object, taking the union of all dates:

```
> empty <- zoo(,dates)            # 'dates' is vector of the missing dates
> merge(ts, empty, all=TRUE)
```

Discussion

The zoo package includes a handy feature in the constructor for zoo objects: you can omit the data and build a *zero-width* object. The object contains no data, just dates. We can use these "Frankenstein" objects to perform such operations as filling and padding on other time series objects.

Suppose you download monthly CPI data for the 1970s. The data is timestamped with the first day of each month:

```
> cpi
1970-01-01 1970-02-01 1970-03-01 1970-04-01 1970-05-01 1970-06-01
      37.8       38.0       38.2       38.5       38.6       38.8
.
. (etc.)
.
1979-07-01 1979-08-01 1979-09-01 1979-10-01 1979-11-01 1979-12-01
      73.1       73.8       74.6       75.2       75.9       76.7
```

As far as R knows, we have no observations for the other days of the months. However, we know that each CPI value applies to the subsequent days through month-end. So first we build a zero-width object with every day of the decade, but no data:

```
> dates <- seq(from=as.Date("1970-01-01"), to=as.Date("1979-12-31"), by=1)
> empty <- zoo(,dates)
```

Then we take the union the CPI data and the zero-width object, yielding a dataset filled with NA values:

```
> filled.cpi <- merge(cpi, empty, all=TRUE)
> filled.cpi
1970-01-01 1970-01-02 1970-01-03 1970-01-04 1970-01-05 1970-01-06 1970-01-07
      37.8         NA         NA         NA         NA         NA         NA
1970-01-08 1970-01-09 1970-01-10 1970-01-11 1970-01-12 1970-01-13 1970-01-14
        NA         NA         NA         NA         NA         NA         NA
1970-01-15 1970-01-16 1970-01-17 1970-01-18 1970-01-19 1970-01-20 1970-01-21
        NA         NA         NA         NA         NA         NA         NA
1970-01-22 1970-01-23 1970-01-24 1970-01-25 1970-01-26 1970-01-27 1970-01-28
        NA         NA         NA         NA         NA         NA         NA
1970-01-29 1970-01-30 1970-01-31 1970-02-01 1970-02-02 1970-02-03 1970-02-04
        NA         NA         NA       38.0         NA         NA         NA
.
. (etc.)
.
```

The resulting time series contains every calendar day, with NAs where there was no observation. That might be what you need. However, a more common requirement is to replace each NA with the most recent observation as of that date. The na.locf function from the zoo package does exactly that:

```
> filled.cpi <- na.locf(merge(cpi, empty, all=TRUE))
> filled.cpi
1970-01-01 1970-01-02 1970-01-03 1970-01-04 1970-01-05 1970-01-06 1970-01-07
      37.8       37.8       37.8       37.8       37.8       37.8       37.8
1970-01-08 1970-01-09 1970-01-10 1970-01-11 1970-01-12 1970-01-13 1970-01-14
      37.8       37.8       37.8       37.8       37.8       37.8       37.8
1970-01-15 1970-01-16 1970-01-17 1970-01-18 1970-01-19 1970-01-20 1970-01-21
      37.8       37.8       37.8       37.8       37.8       37.8       37.8
1970-01-22 1970-01-23 1970-01-24 1970-01-25 1970-01-26 1970-01-27 1970-01-28
      37.8       37.8       37.8       37.8       37.8       37.8       37.8
1970-01-29 1970-01-30 1970-01-31 1970-02-01 1970-02-02 1970-02-03 1970-02-04
      37.8       37.8       37.8       38.0       38.0       38.0       38.0
.
. (etc.)
.
```

January's value of 37.8 is carried forward until February 1, at which time it is replaced by the February value of 38.0. Now every day has the latest CPI value as of that date.

We can use this recipe to fix the problem mentioned in Recipe 14.5. There, the daily price of IBM stock and the monthly CPI data had no intersection on certain days. We can fix that using several different methods. One way is to pad the IBM data to include the CPI dates and then take the intersection (recall that index(cpi) returns all the dates in the CPI time series):

```
> filled.ibm <- na.locf(merge(ibm, zoo(,index(cpi))))
> merge(filled.ibm, cpi, all=FALSE)
           filled.ibm  cpi
1970-02-01     335.19 38.0
1970-03-01     340.25 38.2
1970-04-01     337.75 38.5
1970-05-01     296.75 38.6
1970-06-01     287.00 38.8
1970-07-01     254.00 39.0
.
. (etc.)
.
```

That gives monthly observations. Another way is to fill out the CPI data (as described previously) and then take the intersection with the IBM data. That gives daily observations, as follows:

```
> merge(ibm, filled.cpi, all=FALSE)
                  ibm filled.cpi
1970-01-02 364.75        37.8
1970-01-05 368.25        37.8
1970-01-06 368.50        37.8
1970-01-07 368.75        37.8
1970-01-08 369.50        37.8
1970-01-09 369.00        37.8
.
. (etc.)
.
```

14.7 Lagging a Time Series

Problem

You want to shift a time series in time, either forward or backward.

Solution

Use the lag function. The second argument, k, is the number of periods to shift the data:

```
> lag(ts, k)
```

Use positive *k* to shift the data forward in time (tomorrow's data becomes today's data). Use a negative *k* to shift the data backward in time (yesterday's data becomes today's data).

Discussion

Recall the five days of IBM stock prices from Recipe 14.1:

```
> ibm.daily
2010-01-04 2010-01-05 2010-01-06 2010-01-07 2010-01-08
    132.45     130.85     130.00     129.55     130.85
```

To shift the data forward one day, we use k=+1:

```
> lag(ibm.daily, k=+1, na.pad=TRUE)
2010-01-04 2010-01-05 2010-01-06 2010-01-07 2010-01-08
    130.85     130.00     129.55     130.85         NA
```

We also set na.pad=TRUE to fill the trailing dates with NA. Otherwise, they would simply be dropped, resulting in a shortened time series.

To shift the data backward one day, we use k=-1. Again we use na.pad=TRUE to pad the beginning with NAs:

```
> lag(ibm.daily, k=-1, na.pad=TRUE)
2010-01-04 2010-01-05 2010-01-06 2010-01-07 2010-01-08
        NA     132.45     130.85     130.00     129.55
```

If the sign convention for *k* seems odd to you, you are not alone.

 The function is called lag, but a positive *k* actually generates *leading* data, not lagging data. Use a negative *k* to get *lagging* data. Yes, this is bizarre. Perhaps the function should have been called lead.

14.8 Computing Successive Differences

Problem

Given a time series, *x*, you want to compute the difference between successive observations: $(x_2 - x_1)$, $(x_3 - x_2)$, $(x_4 - x_3)$,

Solution

Use the diff function:

```
> diff(x)
```

Discussion

The `diff` function is generic, so it works on both simple vectors and zoo objects. The beauty of differencing a zoo object is that the result is also a zoo object *and* the differences have the correct dates. Here we compute the differences for successive prices of IBM stock:

```
> ibm.daily
2010-01-04 2010-01-05 2010-01-06 2010-01-07 2010-01-08
    132.45     130.85     130.00     129.55     130.85
> diff(ibm.daily)
2010-01-05 2010-01-06 2010-01-07 2010-01-08
     -1.60      -0.85      -0.45       1.30
```

The difference labeled 2010-01-05 is the change from the previous day (2010-01-04), which is usually what you want. The differenced series is shorter than the original series by one element because R can't compute the change as of 2010-01-04, of course.

By default, `diff` computes successive differences. You can compute differences that are more widely spaced by using its `lag` parameter. Suppose you have monthly CPI data and want to compute the change from the previous 12 months, giving the year-over-year change. Specify a lag of 12:

```
> head(cpi, 24)
1970-01-01 1970-02-01 1970-03-01 1970-04-01 1970-05-01 1970-06-01 1970-07-01
      37.8       38.0       38.2       38.5       38.6       38.8       39.0
1970-08-01 1970-09-01 1970-10-01 1970-11-01 1970-12-01 1971-01-01 1971-02-01
      39.0       39.2       39.4       39.6       39.8       39.8       39.9
1971-03-01 1971-04-01 1971-05-01 1971-06-01 1971-07-01 1971-08-01 1971-09-01
      40.0       40.1       40.3       40.6       40.7       40.8       40.8
1971-10-01 1971-11-01 1971-12-01
      40.9       40.9       41.1
> diff(cpi, lag=12)            # Compute year-over-year change
1971-01-01 1971-02-01 1971-03-01 1971-04-01 1971-05-01 1971-06-01 1971-07-01
       2.0        1.9        1.8        1.6        1.7        1.8        1.7
1971-08-01 1971-09-01 1971-10-01 1971-11-01 1971-12-01 1972-01-01 1972-02-01
       1.8        1.6        1.5        1.3        1.3        1.3        1.4
.
. (etc.)
.
```

14.9 Performing Calculations on Time Series

Problem

You want to use arithmetic and common functions on time series data.

Solution

No problem. R is pretty clever about operations on zoo objects. You can use arithmetic operators (+, -, *, /, etc.) as well as common functions (sqrt, log, etc) as well as usually get what you expect.

Discussion

When you perform arithmetic on zoo objects, R aligns the objects according to date so that the results make sense. Suppose we want to compute the percentage change in IBM stock. We need to divide the daily change by the price, but those two time series are not naturally aligned—they have different start times and different lengths:

```
> ibm.daily
2010-01-04 2010-01-05 2010-01-06 2010-01-07 2010-01-08
    132.45     130.85     130.00     129.55     130.85
> diff(ibm.daily)
2010-01-05 2010-01-06 2010-01-07 2010-01-08
    -1.60      -0.85      -0.45       1.30
```

Fortunately, when we divide one series by the other, R aligns the series for us and returns a zoo object:

```
> diff(ibm.daily) / ibm.daily
  2010-01-05    2010-01-06    2010-01-07    2010-01-08
-0.012227742 -0.006538462 -0.003473562  0.009935040
```

We can scale the result by 100 to compute the percentage change, and the result is another a zoo object:

```
> 100 * (diff(ibm.daily) / ibm.daily)
2010-01-05 2010-01-06 2010-01-07 2010-01-08
-1.2227742 -0.6538462 -0.3473562  0.9935040
```

Functions work just as well. If we compute the logarithm or square root of a zoo object, the result is a zoo object with the timestamps preserved:

```
> log(ibm.daily)
2010-01-04 2010-01-05 2010-01-06 2010-01-07 2010-01-08
  4.886205   4.874052   4.867534   4.864067   4.874052
```

In investment management, computing the difference of logarithms of prices is quite common. That's a piece of cake in R:

```
> diff(log(ibm.daily))
  2010-01-05    2010-01-06    2010-01-07    2010-01-08
-0.012153587 -0.006517179 -0.003467543  0.009984722
```

See Also

See Recipe 14.8 for the special case of computing successive differences.

14.10 Computing a Moving Average

Problem

You want to compute the moving average of a time series.

Solution

Use the `rollmean` function of the `zoo` package to calculate the *k*-period moving average:

```
> library(zoo)
> ma <- rollmean(ts, k)
```

Here *ts* is the time series data, captured in a `zoo` object, and *k* is the number of periods.

For most financial applications, you want `rollmean` to calculate the mean using only historical data; that is, for each day, use only the data available that day. To do that, specify `align=right`. Otherwise, `rollmean` will "cheat" and use future data that was actually unavailable at the time:

```
> ma <- rollmean(ts, k, align="right")
```

Discussion

Traders are fond of moving averages for smoothing out fluctuations in prices. The formal name is the *rolling mean*. You could calculate the rolling mean as described in Recipe 14.12, by combining the `rollapply` function and the `mean` function, but `rollmean` is much faster.

Beside speed, the beauty of `rollmean` is that it returns a `zoo` object. For each element in the object, its date is the "as of" date for a calculated mean. Because it's a `zoo` object, you can easily merge the original data and the moving average and then plot them together.

The output is normally missing a few initial data points, since `rollmean` needs a full *k* observations to compute the mean. Consequently, the output is shorter than the input. If that's a problem, specify `na.pad=TRUE`; then `rollmean` will pad the initial output with NA values.

See Also

See Recipe 14.12 for more about the `align` parameter.

The moving average described here is a *simple moving average*. The quantmod, TTR, and fTrading packages contain functions for computing and plotting many kinds of moving averages, including simple ones.

14.11 Applying a Function by Calendar Period

Problem

Given a time series, you want to group the contents by a calendar period (e.g., week, month, or year) and then apply a function to each group.

Solution

The xts package includes functions for processing a time series by day, week, month, quarter, or year:

```
> apply.daily(ts, f)
> apply.weekly(ts, f)
> apply.monthly(ts, f)
> apply.quarterly(ts, f)
> apply.yearly(ts, f)
```

Here *ts* is an xts time series and *f* is the function to apply to each day, week, month, quarter, or year.

If your time series is a zoo object, convert to an xts object first so you can access these functions; for example:

```
> apply.monthly(as.xts(ts), f)
```

Discussion

It is common to process time series data according to calendar period. But figuring calendar periods is tedious at best and bizarre at worst. Let these functions do the heavy lifting.

Suppose we have a five-year history of IBM stock prices stored in a zoo object:

```
> head(ibm)
1970-01-02 1970-01-05 1970-01-06 1970-01-07 1970-01-08 1970-01-09
    364.75     368.25     368.50     368.75     369.50     369.00
```

We can calculate the average price by month if we use apply.monthly and mean together:

```
> apply.monthly(as.xts(ibm), mean)
                [,1]
1970-01-30 359.8843
1970-02-27 346.8684
1970-03-31 327.4348
.
. (etc.)
.
1974-10-31 176.4791
1974-11-29 181.0570
1974-12-31 167.9233
```

Notice that I first converted the zoo object to an xts object via `as.xts(ibm)`. This would be unnecessary had I stored the data in an xts object in the first place.

A more interesting application is calculating volatility by calendar month, where volatility is measured as the standard deviation of daily log-returns. Daily log-returns are calculated this way:

```
diff(log(ibm))
```

We calculate their standard deviation, month by month, like this:

```
apply.monthly(as.xts(diff(log(ibm))), sd)
```

We can scale the daily number to estimate annualized volatility:

```
sqrt(251) * apply.monthly(as.xts(diff(log(ibm))), sd)
```

Plotting the result produces the chart shown in Figure 14-3, which was created by this code snippet:

```
> plot(sqrt(251) * apply.monthly(as.xts(diff(log(ibm))), sd),
+       main="IBM: Monthly Volatility",
+       ylab="Std dev, annualized")
```

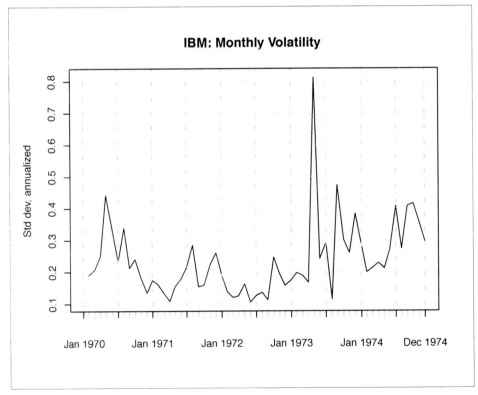

Figure 14-3. IBM volatility by month

14.12 Applying a Rolling Function

Problem

You want to apply a function to a time series in a rolling manner: calculate the function at a data point, move to the next data point, calculate the function at that point, move to the next data point, and so forth.

Solution

Use the `rollapply` function in the zoo package. The *width* parameter defines how many data points from the time series (`ts`) should be processed by the function (*f*) at each point:

```
> library(zoo)
> rollapply(ts, width, f)
```

For many applications, you will likely set `align="right"` to avoid computing *f* with historical data that was unavailable at the time:

```
> rollapply(ts, width, f, align="right")
```

Discussion

The `rollapply` function extracts a "window" of data from your time series, calls your function with that data, saves the result, and moves to the next window—and repeats this pattern for the entire input. As an illustration, consider calling `rollapply` with `width=21`:

```
> rollapply(ts, 21, f)
```

`rollapply` will repeatedly call the function, `f`, with a sliding window of data, like this:

1. `f(ts[1:21])`
2. `f(ts[2:22])`
3. `f(ts[3:23])`
4. ... etc. ...

Observe that the function should expect one argument, which is a vector of values. `rollapply` will save the returned values before packaging them into a zoo object along with a timestamp for every value. The choice of timestamp depends on the `align` parameter given to `rollapply`:

`align="right"`
> The timestamp is taken from the rightmost value.

`align="left"`
> The timestamp is taken from the leftmost value.

```
align="center" (default)
```
The timestamp is taken from the middle value.

By default, rollapply will recalculate the function at successive data points. You may instead want to calculate the function at every *n*th data point. Use the by=*n* parameter to have rollapply move ahead *n* points after each function call. When I calculate the rolling standard deviation of a time series, for example, I usually want each window of data to be separate, not overlapping, and so I set the by value equal to the window size:

```
> sds <- rollapply(ts, 30, sd, by=30, align="right")
```

rollapply normally skips over the initial data points because it needs a full window of data before calling your function. The result is that the output is shorter than the input. You can force them to be the same length by setting na.pad=TRUE, in which case the output will contain NA values for that initial segment.

14.13 Plotting the Autocorrelation Function

Problem

You want to plot the autocorrelation function (ACF) of your time series.

Solution

Use the acf function:

```
> acf(ts)
```

Discussion

The autocorrelation function is an important tool for revealing the interrelationships within a time series. It is a collection of correlations, ρ_k for k = 1, 2, 3, ..., where ρ_k is the correlation between all pairs of data points that are exactly k steps apart.

Visualizing the autocorrelations is much more useful than listing them, so the acf function plots them for each value of k. Figure 14-4 shows the autocorrelation functions for two time series, one with autocorrelations and one without. The dashed line delimits the significant and insignificant correlations: values above the line are significant. (The height of the dashed line is determined by the amount of data.)

The presence of autocorrelations is one indication that an autoregressive integrated moving average (ARIMA) model could model the time series. From the ACF, you can count the number of significant autocorrelations, which is a useful estimate of the number of moving average (MA) coefficients in the model. The top graph in Figure 14-4 shows seven significant autocorrelations, for example, so we estimate that its ARIMA model will require seven MA coefficients (MA(7)). That estimate is just a starting point, however, and must be verified by fitting and diagnosing the model.

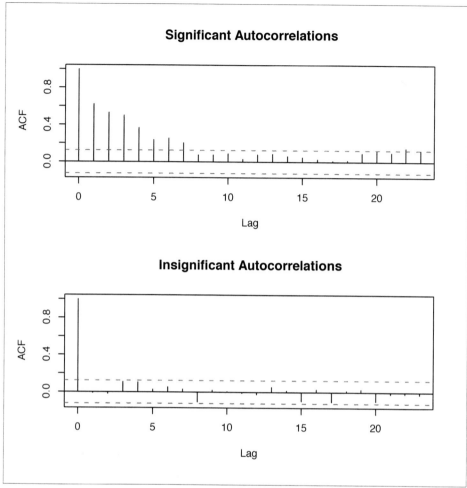

Figure 14-4. Two autocorrelation functions

14.14 Testing a Time Series for Autocorrelation

Problem

You want to test your time series for the presence of autocorrelations.

Solution

Use the Box.test function, which implements the Box–Pierce test for autocorrelation:

```
> Box.test(ts)
```

The output includes a *p*-value. Conventionally, a *p*-value of less than 0.05 indicates that the data contains significant autocorrelations, whereas a *p*-value exceeding 0.05 provides no such evidence.

Discussion

Graphing the autocorrelation function is useful for digging into your data. Sometimes, however, we just need to know whether or not the data is autocorrelated. A statistical test such as the Box–Pierce test can provide an answer.

We can apply the Box–Pierce test to the data whose autocorrelation function we plotted in Recipe 14.13. The test shows *p*-values for the two time series that are nearly zero and 0.862, respectively:

```
> Box.test(ts1)

        Box-Pierce test

data:  ts1
X-squared = 96.9396, df = 1, p-value < 2.2e-16

> Box.test(ts2)

        Box-Pierce test

data:  ts2
X-squared = 0.0303, df = 1, p-value = 0.862
```

The *p*-value near zero indicates that the first time series has significant autocorrelations. (We don't know which autocorrelations are significant; we just know they exist.) The *p*-value of 0.862 indicates that the test did not detect autocorrelations in second time series has none.

The Box.test function can also perform the Ljung–Box test, which is better for small samples. That test calculates a *p*-value whose interpretation is the same as that for the Box–Pierce *p*-value:

```
> Box.test(ts, type="Ljung-Box")
```

See Also

See Recipe 14.13 to plot the autocorrelation function, a visual autocorrelation check.

14.15 Plotting the Partial Autocorrelation Function

Problem

You want to plot the partial autocorrelation function (PACF) for your time series.

Solution

Use the pacf function:

```
> pacf(ts)
```

Discussion

The partial autocorrelation function is another tool for revealing the interrelationships in a time series. However, its interpretation is much less intuitive than that of the autocorrelation function. I'll leave the mathematical definition of partial correlation to a textbook on statistics. Here, I'll just say that the *partial correlation* between two random variables, X and Y, is the correlation that remains after accounting for the correlation shown by X and Y, with all other variables. In the case of time series, the *partial autocorrelation at lag k* is the correlation between all data points that are exactly k steps apart—after accounting for their correlation with the data between those k steps.

The practical value of a PACF is that it helps you to identify the number of autoregression (AR) coefficients in an ARIMA model. Figure 14-5 shows the PACF for two time series, one with partial autocorrelations and one without. Lag values whose lines cross above the dotted line are statistically significant. There are two such values, at $k = 1$ and $k = 2$, so our initial ARIMA model will have two AR coefficients (AR(2)). As with autocorrelation, however, that is just an initial estimate and must verified by fitting and diagnosing the model.

See Also

See Recipe 14.13.

14.16 Finding Lagged Correlations Between Two Time Series

Problem

You have two time series, and you are wondering if there is a lagged correlation between them.

Solution

Use the ccf function to plot the cross-correlation function, which will reveal lagged correlations:

```
> ccf(ts1, ts2)
```

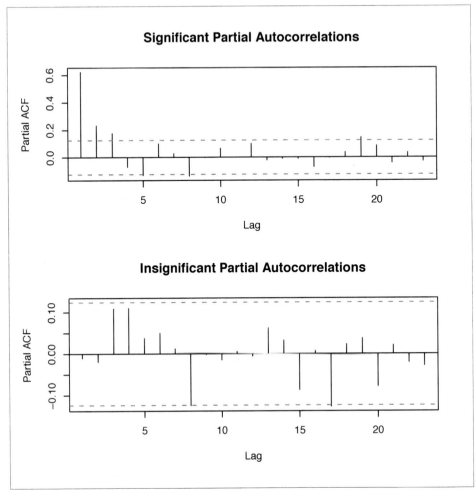

Figure 14-5. Two partial autocorrelation functions

Discussion

The cross-correlation function helps you discover lagged correlations between two time series. A *lagged correlation* occurs when today's value in one time series is correlated with a future or past value in the other time series.

Consider the relationship between commodity prices and bond prices. Some analysts believe those prices are connected because changes in commodity prices are a barometer of inflation, one of the key factors in bond pricing. Can we discover a correlation between them?

Figure 14-6 shows a cross-correlation function generated from daily changes in bond prices and a commodity price index:[†]

```
> ccf(bonds, cmdtys, main="Bonds vs. Commodities")
```

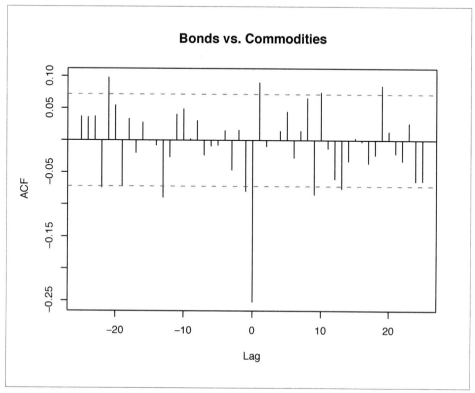

Figure 14-6. Cross-correlation function

Every vertical line shows the correlation between the two time series at some lag, as indicated along the x-axis. If a correlation extends above or below the dotted lines, it is statistically significant.

Notice that the correlation at lag 0 is –0.231, which is the simple correlation between the variables:

```
> cor(bonds, cmdtys)
[1] -0.2309180
```

† Specifically: the bonds variable is the log-returns of the 10-year U.S. Treasury Note futures; and the cmdtys variable is the log-returns of the Rogers International Commodity Index. The data was taken from the period 2007-01-03 through 2009-12-31.

Much more interesting are the correlations at ±1 lag, which are statistically significant. Evidently there is some "ripple effect" in the day-to-day prices of bonds and commodities because changes today are correlated with changes tomorrow. Discovering this sort of relationship is useful to short-term forecasters such as market analysts and bond traders.

14.17 Detrending a Time Series

Problem

Your time series data contains a trend that you want to remove.

Solution

Use linear regression to identify the trend component; then subtract the trend component from the original time series. These two lines show how to detrend the zoo object ts and put the result in detr:

```
> m <- lm(coredata(ts) ~ index(ts))
> detr <- zoo(resid(m), index(ts))
```

Discussion

Some time series data contain trends, which means that they gradually slope upward or downward over time. Suppose your time series object, ts, contains a trend as shown in the upper graph of Figure 14-7.

We can remove the trend component in two easy steps. First, identify the overall trend by using the linear model function, lm. The model should use the time series index for the x variable and the time series data for the y variable:

```
> m <- lm(coredata(ts) ~ index(ts))
```

Second, remove the linear trend from the original data by subtracting the straight line found by lm. This is easy because we have access to the linear model's *residuals*, which are defined by the difference between the original data and the fitted line:

$$r_i = y_i - \beta_1 x_i - \beta_0$$

where r_i is the ith residual and where β_1 and β_0 are the model's slope and intercept, respectively. We can extract the residuals from the linear model by using the resid function and then embed the residuals inside a zoo object:

```
> detr <- zoo(resid(m), index(ts))
```

Notice that we use the same time index as the original data. When we plot detr it is clearly trendless, as evident in the lower graph of Figure 14-7.

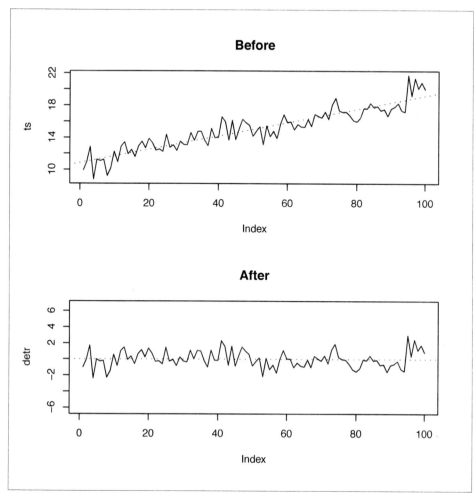

Figure 14-7. Detrending a time series

14.18 Fitting an ARIMA Model

Problem

You want to fit an ARIMA model to your time series data.

Solution

The `auto.arima` function in the `forecast` package can select the correct model order
and fit the model to your data:

```
> library(forecast)
> auto.arima(x)
```

If you already know the model order, (p, d, q), then the arima function can fit the model directly:

```
> arima(x, order=c(p,d,q))
```

Discussion

Creating an ARIMA model involves three steps:

1. Identify the model order.
2. Fit the model to the data, giving the coefficients.
3. Apply diagnostic measures to validate the model.

The model *order* is usually denoted by three integers, (p, d, q), where p is the number of autoregressive coefficients; d is the degree of differencing; and q is the number of moving average coefficients.

When I build an ARIMA model, I am usually clueless about the appropriate order. Rather than tediously searching for the best combination of p, d, and q, I use auto.arima, which does the searching for me:

```
> auto.arima(x)
Series: x
ARIMA(2,1,2)

Call: auto.arima(x = x)

Coefficients:
         ar1      ar2      ma1     ma2
      0.0451  -0.9183  -0.0066  0.6178
s.e.  0.0249   0.0252   0.0496  0.0517

sigma^2 estimated as 0.9877:  log likelihood = -708.45
AIC = 1426.9    AICc = 1427.02    BIC = 1447.98
```

In this case, auto.arima decided the best order was (2, 1, 2), which means that it differenced the data once (d = 1) before selecting a model with two AR coefficients (p = 2) and two MA coefficients (q = 2).

By default, auto.arima limits p and q to the range $0 \leq p \leq 5$ and $0 \leq q \leq 5$. If you are confident that your model needs fewer than five coefficients, use the max.p and max.q parameters to limit the search further; this makes it faster. Likewise, if you believe that your model needs more coefficients, use max.p and max.q to expand the search limits.

If you already know the order of your ARIMA model, the arima function can quickly fit the model to your data:

```
> arima(x, order=c(2,1,2))
Series: x
ARIMA(2,1,2)

Call: arima(x = x, order = c(2, 1, 2))
```

```
Coefficients:
         ar1      ar2      ma1     ma2
      0.0451  -0.9183  -0.0066  0.6178
s.e.  0.0249   0.0252   0.0496  0.0517

sigma^2 estimated as 0.9877:  log likelihood = -708.45
AIC = 1426.9    AICc = 1427.02   BIC = 1447.98
```

The output looks identical to `auto.arima`. What you can't see here is that `arima` executes much more quickly.

The output from `auto.arima` and `arima` includes the fitted coefficients and the standard error (`s.e.`) for each coefficient:

```
Coefficients:
         ar1      ar2      ma1     ma2
      0.0451  -0.9183  -0.0066  0.6178
s.e.  0.0249   0.0252   0.0496  0.0517
```

You can find the coefficients' confidence intervals by capturing the ARIMA model in an object and then using the `confint` function:

```
> m <- arima(x, order=c(2,1,2))
> confint(m)
             2.5 %        97.5 %
ar1  -0.003811699   0.09396993
ar2  -0.967741073  -0.86891013
ma1  -0.103861019   0.09059546
ma2   0.516499037   0.71903424
```

This output illustrates a major headache of ARIMA modeling: not all the coefficients are necessarily significant. The `ar1` coefficient has a confidence interval of (–0.0038, +0.0940). Since the interval contains zero, the true coefficient might be zero itself, in which case the whole `ar1` term is unnecessary. The same is true for the `ma1` coefficient, whose confidence interval is (–0.0066, +0.0496). The `ar2` and `ma2` coefficients are likely significant, however, since their confidence intervals do *not* contain zero.

If you discover that your model contains insignificant coefficients, use Recipe 14.19 to remove them.

The `auto.arima` and `arima` functions contain useful features for fitting the best model. For example, you can force them to include or exclude a trend component. See the help pages for details.

A final caveat: the danger of `auto.arima` is that it makes ARIMA modeling look simple. ARIMA modeling is *not* simple. It is more art than science, and the automatically generated model is just a starting point. I urge the reader to review a good book about ARIMA modeling before settling on a final model.

See Also

See Recipe 14.20 for performing diagnostic tests on the ARIMA model.

14.19 Removing Insignificant ARIMA Coefficients

Problem

One or more of the coefficients in your ARIMA model are statistically insignificant. You want to remove them.

Solution

The `arima` function includes the parameter `fixed`, which is a vector. The vector should contain one element for every coefficient in the model, including a term for the drift (if any). Each element is either NA or 0. Use NA for the coefficients to be kept and use 0 for the coefficients to be removed. This example shows an ARIMA(2, 1, 2) model with the first AR coefficient and the first MA coefficient forced to be 0:

```
> arima(x, order=c(2,1,2), fixed=c(0,NA,0,NA))
```

Discussion

The example in Recipe 14.18 contained two insignificant coefficients, `ar1` and `ma1`. We decided they were insignificant by looking at the confidence intervals:

```
> m <- arima(x, order=c(2,1,2))
> confint(m)
            2.5 %      97.5 %
ar1 -0.003811699   0.09396993
ar2 -0.967741073  -0.86891013
ma1 -0.103861019   0.09059546
ma2  0.516499037   0.71903424
```

Since the confidence intervals for `ar1` and `ma1` contain zero, the true coefficients could be zero. Those are the first and third coefficients, so we can remove them by letting `fixed` be a vector whose first and third elements are zero:

```
> m <- arima(x, order=c(2,1,2), fixed=c(0,NA,0,NA))
> m
Series: x
ARIMA(2,1,2)

Call: arima(x = x, order = c(2, 1, 2), fixed = c(0, NA, 0, NA))

Coefficients:
      ar1     ar2  ma1     ma2
        0  -0.9082    0  0.5931
s.e.    0   0.0268    0  0.0522

sigma^2 estimated as 0.999:  log likelihood = -711.27
AIC = 1428.54    AICc = 1428.66    BIC = 1449.62
Warning message:
In arima(x, order = c(2, 1, 2), fixed = c(0, NA, 0, NA)) :
  some AR parameters were fixed: setting transform.pars = FALSE
```

The warning message from `arima` is harmless. The default value for the `transform.pars` parameter is `TRUE`, but `arima` forces it to `FALSE`, if necessary, and warns you.

Observe that the `ar1` and `ma1` coefficients are now zero. The remaining coefficients (`ar2` and `ma2`) are still significant, as shown by their confidence intervals, so we have a reasonable model:

```
> confint(m)
          2.5 %      97.5 %
ar1          NA          NA
ar2  -0.9606839  -0.8557283
ma1          NA          NA
ma2   0.4907792   0.6954606
```

14.20 Running Diagnostics on an ARIMA Model

Problem

You have built an ARIMA model, and you want to run diagnostic tests to validate the model.

Solution

Use the `tsdiag` function. This example saves the ARIMA model in `m` and then runs diagnostics on the model:

```
> m <- arima(x, order=c(2,1,2), fixed=c(0,NA,0,NA), transform.pars=FALSE)
> tsdiag(m)
```

Discussion

The result of `tsdiag` is a set of three graphs. A good model should produce results like the ones shown in Figure 14-8.

Here's what's good about the graphs:

- The standardized residuals don't show clusters of volatility.
- The autocorrelation function (ACF) shows no significant autocorrelation between the residuals.
- The *p*-values for the Ljung–Box statistics are all large, indicating that the residuals are patternless. We have extracted the information and left behind only "noise."

For contrast, Figure 14-9 shows diagnostic charts with problems:

- The ACF shows significant autocorrelations between residuals.
- The *p*-values for the Ljung–Box statistics are small, indicating there is some pattern in the residuals. There is still information to be extracted from the data.

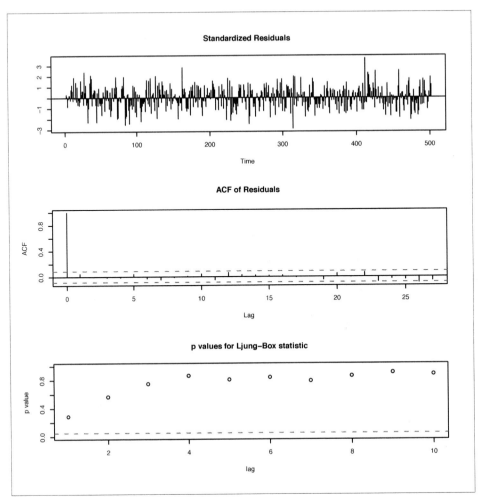

Figure 14-8. Good time series diagnostics

These are basic diagnostics, but they are a good start. Find a good book on ARIMA modeling and perform the recommended diagnostic tests before concluding that your model is sound. Additional checks of the residuals could include:

- Tests for normality
- Quantile-quantile (Q-Q) plot
- Histogram
- Scatterplot against the fitted values
- Time-series plot

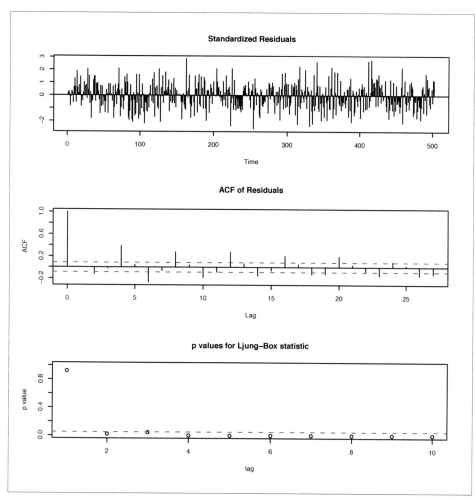

Figure 14-9. Bad time series diagnostics

14.21 Making Forecasts from an ARIMA Model

Problem

You have an ARIMA model for your time series. You want to forecast the next few observations in the series.

Solution

Save the model in an object; then apply the `predict` function to the object. This example saves the model from Recipe 14.19 and predicts the next observation:

```
> m <- arima(x, order=c(2,1,2), fixed=c(0,NA,0,NA), transform.pars=FALSE)
> predict(m)
```

Discussion

The predict function will calculate both the next observation and its standard error according to the model. It returns a list with two elements: pred, which is the prediction itself; and se, which is the standard error. Continuing the example from the Solution, the predicted next value is –8.545 with a standard error of 0.9995:

```
> predict(m)
$pred
Time Series:
Start = 503
End = 503
Frequency = 1
[1] -8.545194

$se
Time Series:
Start = 503
End = 503
Frequency = 1
[1] 0.9995073
```

(Both pred and se are encoded as time series, so R prints their time series attributes in addition to their numeric values.) To predict more than one observation into the future, use the n.ahead parameter:

```
> predict(m, n.ahead=10)
$pred
Time Series:
Start = 503
End = 512
Frequency = 1
 [1] -8.545194 -9.365685 -9.365838 -8.620663 -8.620524 -9.297297 -9.297423
 [8] -8.682773 -8.682659 -9.240887

$se
Time Series:
Start = 503
End = 512
Frequency = 1
 [1] 0.9995073 1.4135168 1.5705650 1.7132774 1.9691066 2.1953229 2.3075363
 [8] 2.4145404 2.5935203 2.7609219
```

As before, the output includes two time series: one of predictions and one of standard errors. Notice that the standard errors grow larger with each additional prediction because the predictions become increasingly more uncertain.

14.22 Testing for Mean Reversion

Problem

You want to know if your time series is mean reverting.

Solution

A common test for mean reversion is the Augmented Dickey–Fuller (ADF) test, which is implemented by the `adf.test` function of the `tseries` package:

```
> library(tseries)
> adf.test(coredata(ts))
```

The output from `adf.test` includes a p-value. Conventionally, if $p < 0.05$ then the time series is likely mean reverting whereas a $p > 0.05$ provides no such evidence.

Discussion

When a time series is *mean reverting*, it tends to return to its long-run average. It may wander off, but eventually it wanders back. If a time series is not mean reverting, then it can wander away without ever returning to the mean.

The upper time series in Figure 14-10 appears to be wandering downward and not returning. The large p-value from the `adf.test` confirms that it is not mean reverting:

```
> adf.test(coredata(ts1))

        Augmented Dickey-Fuller Test

data:  ts1
Dickey-Fuller = -2.1217, Lag order = 6, p-value = 0.5246
alternative hypothesis: stationary
```

The lower time series, however, is just bouncing around its average value. The small p-value (0.01) confirms that it is mean reverting:

```
> adf.test(coredata(ts2))

        Augmented Dickey-Fuller Test

data:  ts2
Dickey-Fuller = -7.1752, Lag order = 6, p-value = 0.01
alternative hypothesis: stationary
```

The `adf.test` function massages your data before performing the ADF test. First it automatically detrends your data; then it recenters the data, giving it a mean of zero.

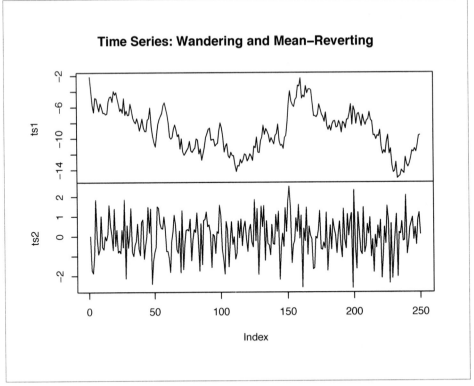

Figure 14-10. Two time series: wandering and mean reverting

If either detrending or recentering is undesirable for your application, use the `adfTest` function in the fUnitRoots package instead:‡

```
> library(fUnitRoots)
> adfTest(coredata(ts1), type="nc")
```

With type="nc", the function neither detrends nor recenters your data. With type="c", the function recenters your data but does not detrend it.

Both the `adf.test` and the `adfTest` functions let you specify a lag value that controls the exact statistic they calculate. These functions provide reasonable defaults, but any serious user should study the textbook description of the ADF test to determine the appropriate lag for her application.

‡ Make careful note of the difference in spelling between `adf.test` and `adfTest`. This sort of subtle difference is unfortunately common in R and occurs when several packages implement the same function but with some slight variation from package to package.

See Also

The urca and CADFtest packages also implement tests for a unit root, which is the test for mean reversion. Be careful when comparing the tests from several packages. Each package can make slightly different assumptions, which can lead to puzzling differences in the results.

14.23 Smoothing a Time Series

Problem

You have a noisy time series. You want to smooth the data to eliminate the noise.

Solution

The KernSmooth package contains functions for smoothing. Use the dpill function to select an initial bandwidth parameter; then use the locploy function to smooth the data:

```
library(KernSmooth)

gridsize <- length(y)
bw <- dpill(t, y, gridsize=gridsize)
lp <- locpoly(x=t, y=y, bandwidth=bw, gridsize=gridsize)
smooth <- lp$y
```

Here, t is the time variable and y is the time series.

Discussion

The KernSmooth package is a standard part of the R distribution. It includes the locpoly function; this function constructs, around each data point, a polynomial that is fitted to the near-by data points. These are called *local polynomials*. The local polynomials are strung together to create a smoothed version of the original data series.

The algorithm requires a *bandwidth* parameter to control the degree of smoothing. A small bandwidth means less smoothing, in which case the result follows the original data more closely. A large bandwidth means more smoothing, so the result contains less noise. The tricky part is choosing just the right bandwidth: not too small, not too large.

Fortunately, KernSmooth also includes the function dpill for estimating the appropriate bandwidth, and it works quite well. I recommend that you start with the dpill value and then experiment with values above and below that starting point. There is no magic formula here. You need to decide what level of smoothing works best in your application.

Figure 14-11 shows an example of smoothing. The solid line is a time series that is the sum of a simple sine wave and normally distributed "noise":

```
t <- seq(from=-10,to=10,length.out=201)
noise <- rnorm(201)
y <- sin(t) + noise
```

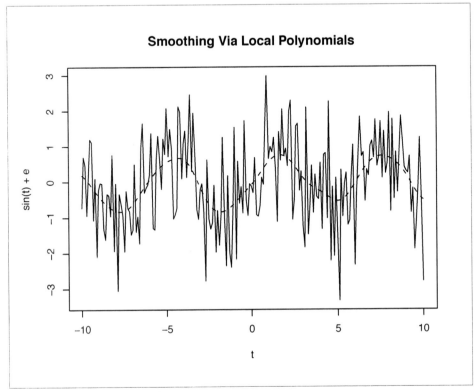

Figure 14-11. Smoothing a time series

Both dpill and locpoly require a grid size—in other words, number of points for which a local polynomial is constructed. I often use a grid size equal to the number of data points, which yields a fine resolution. The resulting time series is very smooth. You might use a smaller grid size if you want a coarser resolution or if you have a very large dataset:

```
library(KernSmooth)
gridsize <- length(y)
bw <- dpill(t, y, gridsize=gridsize)
```

The locpoly function performs the smoothing and returns a list. The y element of that list is the smoothed data:

```
lp <- locpoly(x=t, y=y, bandwidth=bw, gridsize=gridsize)
smooth <- lp$y
```

In Figure 14-11, the smoothed data is shown as a dashed line. The figure demonstrates that `locpoly` did an excellent job of extracting the original sine wave.

See Also

The `ksmooth`, `lowess`, and `HoltWinters` functions in the base distribution can also perform smoothing. The `expsmooth` package implements exponential smoothing.

Index

Symbols

! (logical negation), 40
!= (comparison operator), 34, 40
(comment), 80
#! (Linux/Unix self-contained script), 65
$ (component extraction), 40
%% (modulo operator), 41, 319
%*% (matrix multiplication), 41, 120
%...% (binary operators), 40
%/% (integer division), 41
%in% (contained in operator), 41
& (logical and), 40, 48
&& (short-circuit and), 40, 48
() (print assigned value), 315
* (multiplication operator), 40, 120, 280, 286
*, **, *** (significance codes), 277, 297
+ (continuation prompt), 47
+ (unary plus, addition operator), 40, 286
- (unary minus, subtraction operator), 40, 286
--help option, 6
--no-init option, 64
--no-restore option, 64
--no-save option, 64
--quiet option, 62, 64
--slave option, 64
->, ->> (rightwards assignment operators), 26
. (significance code in regression summary), 277
/ (division operator), 40, 286
: (multiplication in model formula), 280
: (sequence creation operator), 33, 40
::, ::: (access variables in name space), 40
< (comparison operator), 34, 40

<-, <<- (leftwards assignment operators), 25, 40, 46
<<- (global assignment operator), 26, 43
<= (comparison operator), 34, 40
= (alternate assignment operator), 26
== (comparison operator), 34, 40, 47
> (command prompt), 7
> (comparison operator), 34, 40
>= (comparison operator), 34, 40
? (help shortcut), 12, 40
?? (search shortcut), 13
@ (slot extraction), 40
[] (vector indexing), 7, 35, 40
[[]] (list indexing), 40, 110
\ (backslash), 46, 76, 168
\n (newline character), 24
^ (exponentiation, interaction term), 40, 285, 287
| (logical or), 40, 48
|| (short-circuit or), 40, 48
~ (formula), 40, 233, 248, 269, 303

A

abline() function, 222, 231, 245, 255
abs() function, 117
abstract type, 97
access variables in name space (::, :::), 40
accessing
 built-in datasets, 57
 data frame column contents, 141
 functions in a package, 55
acf() function, 299, 376
active window, choosing, 262
ad.test() function, 210
adding

We'd like to hear your suggestions for improving our indexes. Send email to *index@oreilly.com*.

dbDisconnect() function, 89
dbGetQuery() function, 89, 90
dbinom() function, 186
Debian, installing R on, 3
defaultPackages list, 69
deleting
 Delete key, 8
 variables, 27
dendrograms, 342
density functions for probability distributions,
 186, 190
density() function, 250
desktop icon, creating, 5
detach() function, 56, 141
detrending a time series, 382, 392
dev.set() function, 262
deviance() function, 272
dfrm list operators, 127
dgamma() function, 191
dgeom() function, 187
diagnostic checks for regression, 293
dictionary, 96
diff() function, 187, 369
directories
 home, 67
 working, 5, 51
discrete distributions, 177, 186, 190
discrete histograms, 252
dist() function, 342
distribution function, 187
distribution functions for probability
 distributions, 186, 188
division operator (/), 40
dnorm() function, 177, 191, 258
do.call() function, 124, 331, 332
documentation
 packages, 3
 supplied, 1, 10, 13
double atomic type, 143
down arrow, 8
downloading and installing R, 2–4
dpill() function, 393
dpois() function, 187
dput() function, 92
dump() function, 21, 92
dumping variable structure, 327
dwtest() function (Durbin–Watson test), 299
dynamically typed language, 25

E

edit() function, 135
editor window, 43, 135
effects() function, 272
eigen() function, 338
eigenvalues and eigenvectors, 338
elapsed time, 329
End key, 8
environment variables, 66
equals sign
 = (alternate assignment operator), 26
 == (comparison operator), 34, 40, 47
errors
 "cannot open the connection", 76
 common syntax, 46–49
 "could not find function", 48, 56
 "has not been called yet", 222
 "not in a standard unambiguous format",
 170
 "object . . . not found", 47
 suppressing messages, 329
escape character (\), 46, 76, 168
Euclid's algorithm, 42
example() function, 12
Excel worksheet, data frame as, 101
excluding data frame columns by name, 137
expand.grid() function, 321
exponential distributions, 178, 188, 190, 255
exponential smoothing, 395
exponentiation (^), 40, 286, 287
expsmooth package, 395
extracting
 Julian date, 173
 oldest/newest observations, 361
 parts of a date, 174
 substrings, 164

F

F distributions, 178
F statistic, 267, 268, 277
factor analysis/factanal() function, 349
factors
 categorical variables as, 105
 creating, 105
 creating box plot for each level, 247
 creating scatter plot for each level, 233
 defining groups via, 147
 factor() function, 105, 126

Mathematical Statistics with Applications
 (Wackerly), 197
matrices
 applying functions to, 151
 dim attribute and, 325, 326
 graphics window, 260
 initializing, 118
 matrix() function, 118
 multiplication (%*%), 41, 120
 naming rows/columns, 120
 normalized values, 203
 operations, 119
 reshaped from vectors, 98, 119
 selecting rows/columns, 121
 summaries of, 198
 summing rows/columns, 315
max() function (maximum values), 48, 318,
 320
maximizing
 multiparameter function, 336
 single-parameter function, 335
means
 forming confidence interval for, 205
 group, finding differences between, 304
 group, pairwise comparisons between, 218
 mean() function, 30–32, 48, 151, 156, 199,
 202, 322
 reversion to, 391
 of two samples, comparing, 212
 testing for zero, 205
median
 confidence interval for, 206
 median() function, 30–32, 49
memory management, 102
merge() function, 140, 364, 366
merging data frames by common column, 140
methods, 97
min() function, 48, 320
minimal self-contained examples, 20
minimizing
 multiparameter function, 336
 single-parameter function, 335
minimum values, 318, 320
mirror sites, CRAN, 60
mode() function, 96
model formula, 269
model objects
 abline() function, 231
 ANOVA comparisons of, 309

diagnostic plots from, 295
glm() function, 345
identifying influential observations from,
 296
lm() function, 267, 273
from residuals, 293
TukeyHSD() function and, 305
Modern Applied Statistics with S (Venables and
 Ripley), xv
modular tools, 71
modulo operator (%%), 41, 319
Moen, Rick, 21
mondate package, 162
mouse, 8
moving average (MA), 376
multifigure plots, 260
multilevel list structures, 331
multiparameter functions, 336
multiple linear regression, 270
multiplication operator (*), 40, 120, 280
multiplication operator in model formula (:),
 280
multiway ANOVA, 303
Murrell, Paul, xv, 221
my.cnf file, 90
MySQL client software, 90

N
NA values
 filling in, 367
 initializing matrix with, 118
 na.locf() function, 365
 na.omit() function, 136
 na.strings parameter, 79
 removing from data frames, 136
 replacing, 365
 in vectors, 31, 102
Nachtsheim, Christopher, 269
name/value association lists, 112
nchar() function, 163
negative binomial distributions, 178
negative indexes, 36
Neter, John, 269
newline character (\n), 24
noise, eliminating, 393
nonparametric statistics, 214
normal density, 177
normal distributions
 functions, 188

getting help for, 178
names of, 177
qnorm() function, 190
testing for, 209, 252
normal quantiles, 177, 189
normalizing data, 203
null hypothesis, 195, 210, 308
NULL value, 114, 116, 325
numeric atomic type, 144
numeric() function, 126

O

objects
 class of, 97, 355
 mode of, 96
 saving and transporting, 92
 structure of, 326
observations, 123
 computing successive differences between,
 369
 extracting oldest/newest, 313, 361
 filling in dates/times for missing, 366
 forecasting with ARIMA, 389
 identifying influential, 296
 paired, 212, 337
oneway.test() function (one-way ANOVA),
 268, 302
operator precedence, 40, 333
operators, binary, 332
optim() function, 336
optimize() function, 335
options() function, 69, 74, 314
order() function, 323, 324
orthogonal regression, 340
OS X
 exiting from R, 9
 installing packages, 59
 installing R, 2
 interrupting R, 9
 location of .Rprofile, 68
 saving graphics files, 263
 starting R, 4, 6
outer() function, 168
outlier.test() function, 294, 295
output, redirecting to file, 74
overwriting errors involving plots, 259

P

p-values, 195, 201
p.adjust() function, 218
pacf() function (partial autocorrelation
 function), 378
packages
 accessing functions in, 55
 attaching to search list, 58
 datasets and, 57
 for dates and times, 162
 documentation for, 1, 3
 graphical managers, 3
 graphics packages, 221, 223
 help on, 14
 installing and loading, 2, 48, 56, 59, 69
 list of, 10, 19
 preloading, 69
 and search path, 54
 unloading, 56
 viewing list of installed, 58
padding a time series, 366
paired observations, 212
pairwise comparison between group means,
 218
pairwise.t.test() function, 218
par() function, 261, 264
parallel minimum/maximum, 320
parallel vectors, 158
parentheses, use of, 46, 47
partial autocorrelation function (PACF), 378
partial correlation, 379
password security, 90
paste() function, 164, 168
pattern searching, 13, 14
pausing between plots, 259
pbinom() function, 187
PCA (principal component analysis), 338, 350
pchisq() function, 188
Pearson chi-square test, 210
Pearson method, 215
pearson.test() function, 210
performing
 calculations on time series, 370
 linear regression with interaction terms,
 279
 linear regression without intercept, 278
 matrix operations, 119
 multiple linear regression, 270
 one-way ANOVA, 302

testing for normality, 210, 252
quantiles
 converting probabilities to, 189
 interquartile range, 247
 inverting, 202
 qbinom() function, 190
 qexp() function, 255
 qgeom() function, 190
 qnorm() function, 177
 qpois() function, 190
 qt() function, 190, 255
 quantile() function, 201
quartiles, 201

R

R
 console version, 6
 design philosophy, 71
 overview of start-up sequence, 70
 R console, 6
 R Language Definition, xvi
 R_HOME environment variable, 67
 technical aspects of data structures, 95
 website, 10
R^2 (coefficient of determination), 277
R Data Import/Export, 71
R Graphics (Murrell), xv, 221
R in a Nutshell (Adler), xv
rainbow() function, 239
random numbers, generating, 180–183
random permutations of vectors, 185
random samples, generating, 183
random sequences, generating, 184
random variates, 177
range() function, 150, 152
rank() function, 154
Raymond, Eric, 21
rbind() function, 124, 125, 138
.RData file, 5, 9, 52, 64, 70
reading
 CSV files, 80
 files with complex structure, 86–89
 fixed-width records, 77
 from MySQL databases, 89
 read.csv() function, 81, 83, 86
 read.fwf() function, 77
 read.table() function, 78, 83, 86
 read.zoo() function, 358
 readHTMLTable() function, 84

readLines() function, 86
 tabular data files, 78
 from the Web, 83–86
Recycling Rule, 48, 103–105, 118, 139, 173
Red Hat, installing R on, 3
reduced model, 282
regression
 on data subset, 284
 diagnosing a linear, 293
 formula, using expression inside, 285
 logistic, 345
 orthogonal, 340
 on polynomials, 286
 summary, understanding the, 275–278
 on transformed data, 287
regression coefficients, 292
regression line of a scatter plot, 231
regression residuals, 293
relative frequencies, 199
removing
 elements from lists, 114, 116, 117
 insignificant ARIMA coefficients, 386
 NA values from data frames, 136
rep() function (repeated values), 33
replacing substrings, 166
reproducible random numbers, 182
require() function, 48, 56, 69
reshape2() package, 140
residuals
 resid() function, 272, 382
 residuals() function, 272
 statistics, 276, 277
 testing for autocorrelation, 298
resources
 R books, xv
 statistics books, xvi
 on the Web, xv
response variables, 267, 269
return value, 42
reversion to mean, 391
.Rhistory file, 53
RHOME subcommand, 67
right arrow, 8
rightwards assignment operators (->, ->>), 26,
 40
Ripley, Brian, xv, 163
ripple effect, 382
rm() function, 27
RMySQL package, 90

rnorm() function, 177
robust ANOVA, 308
rollapply() function, 375
rollmean() function, 372
row labels, 83, 120
rowSums() function, 315
.Rprofile files, 70
.Rprofile files, 62, 64, 68–70
Rscript program, 64
rseek.org, 16, 18
RSiteSearch() function, 16, 19
Rterm.exe (console version), 6
runif() function, 181
runs.test() function, 211

S

S Programming (Venables and Ripley), xvi
S statistical package, 71
samples
 comparing means of two, 212
 proportion-testing, 207
 sample() function, 184, 185, 186
 size of, 205
 testing for same distribution, 219
 testing mean of, 203
 with/without replacement, 184, 185, 347
sapply() function, 116, 149, 152
Sarkar, Deepayan, xv, 223
SAS dataset, data frame as, 101
saving
 to a graphics file, 263
 objects, 92
 save() function, 92
 save.image() function, 5, 52
 savePlot() function, 263
 without exiting R, 52
scalars, 97, 158
scale() function, 203
scan() function, 83, 86
scatter plots, 223, 231–236
scripts, running, 62–65
sd() function, 30–32, 49
searching
 ?? shortcut, 13
 with built-in search engine, 10
 using search path, 54
 search() function, 54
segments() function, 222, 258
selecting

best regression variables, 281
data frame columns by name, 131
data frame columns by position, 127–130
every nth element of vector, 319
matrix row or column, 121
rows and columns with subset(), 132
vector elements, 35
self-describing data, 80
sep parameter, 79
separator, string, 164
sequences
 creation operator (:), 33, 40
 seq() function, 33, 175
 seq_along() function, 319
set.seed() function, 183
setting
 environment variables, 66
 working directory, 51
sf.test() function, 210
shading, density plot with, 191
shapiro.test() function, 209
Shapiro–Francia test, 210
Shapiro–Wilk test, 210
shifting a time series, 368
short-circuit and (&&), 40, 48
short-circuit or (||), 40, 48
shortcuts
 ? help, 12
 creating, 5
 embedding --quiet option with, 62
 starting R from desktop, 5
side-by-side display, 260
significance codes, 277
significance level, 196
simple linear regression, 267, 269
simplifying results, 149
single-parameter function, 335
sink() function, 75
slope, computing, 340
slot extraction (@), 40
smoothing, 393
snippets, using, 44
sorting data frames by two columns, 324
sos package, 19
source() function, 63
Spearman method, 215
special characters, 168
special interest group (SIG) mailing lists, 21,
 49

computing successive differences between, 369

detrending, 382, 392

finding lagged correlations between, 379

lagging, 368

mean reversion testing, 391

merging, 364

overview, 355

performing calculations on, 370

plotting data, 359

representing data from, 356

rolling mean, 372

smoothing, 393

subsetting, 363

timeDate package, 162

timeSeries package, 358

timestamps, 364, 366, 371, 375

timing your code, 329

title() function, 225

TLS (total least squares), 340

tokens in data files, 86

total least squares (TLS), 340

transformed data

 Box–Cox procedure, 289

 regressing on, 287

transporting objects, 92

trends, removing, 382

ts class, 355

tsdiag() function, 387

tseries package, 211, 391

Tukey, John, 305

TukeyHSD() function, 305

U

Ubuntu R install, 3

unary minus operator (-), 40

unary plus operator (+), 40

unclass() function, 326

undefined operators, 333

uniform distributions, 178

Unix (see Linux/Unix)

unlist() function, 115, 117

unprintable characters, 167

unstack() function, 148

up arrow, 8

urca package, 393

user CPU time, 329

Using R for Introductory Statistics (Verzani), xvi

V

values

 missing, 79

 obtaining last, 53

 predicting from regression model, 300

var() function, 30–32

variables

 categorical (see factors)

 data organized by, 123

 deleting, 27

 environment, 66

 global, 43

 .Last.value, 53

 listing, 26

 local, 43

 low-correlation, removing, 153

 setting (assigning), 25

 stripping attributes from, 325

vcov() function (variance-covariance), 273

vectors

 appending data to, 101

 c() constructor for, 72, 101

 calculating z-scores of, 39

 comparing, 34

 creating, 28

 finding position of value, 318

 flattening lists into, 115, 145

 and grouping factors, 147

 indexing of ([]), 35, 40

 inserting data into, 103

 logical, 117

 versus loops, 147

 and matrices, 98, 119

 and modes, 97

 multiple, combining into vector and factor, 107

 of normalized values, 203

 observations stored in, 124

 parallel, 112

 performing arithmetic on, 38–39

 printing simple, 24

 properties of, 95

 recentering, 39

 rotation, 339

 and scalars, 97

 selecting elements of, 35

 selecting every nth element, 319

 summaries of lists of, 197, 198

 of unequal length, 48, 103–105

Venables, William, xv
vertical lines, adding to plots, 245
Verzani, John, xvi
vignette() function, 16, 358

W

Wackerly, Dennis, 197
warning messages
"argument is not numeric or logical", 331
"In arima . . . some AR parameters were fixed", 386
suppressing, 69, 329
unrecognized escape in a character string, 76
warnings() function, 330
web data, reading, 83–86
week, defined, 362
Weibull distributions, 178
which.max() function, 318
which.min() function, 318
while statements, 48
Wickham, Hadley, xv, 223
wilcox.test() function, 214
Wilcoxon distribution, 178
Wilcoxon–Mann–Whitney test, 214
win.graph() function, 262
window() function, 363
windows, graphic, 260, 262
Windows, Microsoft
"cannot open" file error in, 76
exiting from R, 8
installing packages on, 59
installing R on, 2
interrupting R in, 9
location of .Rprofile, 68
"Save as" for graphics files, 263
starting R, 4–6
with() function, 141
Witmer, Jeff, 88
working directory, 5, 51
workspace
data saved outside of, 92
.RData file, 5, 9, 52, 64, 70
restore options, 64
saving, 9, 52, 64
variables held in, 25
World Series dataset, 88
write.csv() function, 82

X

X Windows system, 66
XML package, 84
xtabs() function, 200
xts package, 355, 356, 373

Z

z-scores, 39, 203
zero intercept, 278
zero-width object, 366
zoo objects, performing calculations on, 371
zoo package, 223, 355, 356
Zoonekynd, Vincent, 341

About the Author

Paul Teetor is a quantitative developer with master's degrees in statistics and computer science. He specializes in analytics and software engineering for investment management, securities trading, and risk management. He works with hedge funds, market makers, and portfolio managers in the greater Chicago area.

Colophon

The animal on the cover of *R Cookbook* is a harpy eagle (*Harpia harpyja*). One of the fifty species of eagle in the world, the harpy eagle is native to the tropical rain forests of Central and South America, and prefers to nest in the upper canopy layer thereof. Both its genus and species names refer to the harpies of ancient Greek mythology—vicious creatures with the face of a woman and the body of an eagle or vulture.

On average, harpy eagles weigh about 18 lbs., are 36 to 40 inches long, and have a wingspan of 6 to 7 feet, though females are consistently larger than males. The plumage of both sexes is identical, however: slate black feathers dominate the animal's top half, while the underside is white or light gray. Light gray-colored heads are accentuated with a double crest of large feathers, which specimens can raise when showing hostility.

Harpy eagles are monogamous, and pairs raise only one chick every two to three years. Females will usually lay two eggs at a time, and after the first hatches, the other is neglected and does not hatch. Though the chick will fledge within 6 months, both parents continue to care for and feed the chick for at least a year. Because of this low rate of population growth, the harpy eagle is particularly susceptible to encroachments on its habitat and losses from human hunting. Throughout its range, the animal's conservation status ranges from threatened to critically endangered.

The cover image is from J. G. Wood's *Animate Creation*. The cover font is Adobe ITC Garamond. The text font is Linotype Birka; the heading font is Adobe Myriad Condensed; and the code font is LucasFont's TheSansMonoCondensed.

Get even more for your money.

Join the O'Reilly Community, and register the O'Reilly books you own. It's free, and you'll get:

- $4.99 ebook upgrade offer
- 40% upgrade offer on O'Reilly print books
- Membership discounts on books and events
- Free lifetime updates to ebooks and videos
- Multiple ebook formats, DRM FREE
- Participation in the O'Reilly community
- Newsletters
- Account management
- 100% Satisfaction Guarantee

Signing up is easy:

1. **Go to: oreilly.com/go/register**
2. **Create an O'Reilly login.**
3. **Provide your address.**
4. **Register your books.**

Note: English-language books only

To order books online:
oreilly.com/store

For questions about products or an order:
orders@oreilly.com

To sign up to get topic-specific email announcements and/or news about upcoming books, conferences, special offers, and new technologies:
elists@oreilly.com

For technical questions about book content:
booktech@oreilly.com

To submit new book proposals to our editors:
proposals@oreilly.com

O'Reilly books are available in multiple DRM-free ebook formats. For more information:
oreilly.com/ebooks

Spreading the knowledge of innovators oreilly.com

Buy this book and get access to the online edition for 45 days—for free!

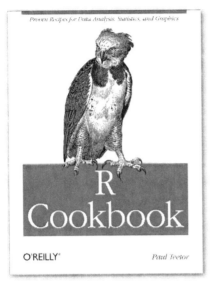

Proven Recipes for Data Analysis, Statistics, and Graphics

R Cookbook

O'REILLY® Paul Teetor

R Cookbook
By Paul Teetor
February 2011, $39.99
ISBN 9780596809157

With Safari Books Online, you can:

Access the contents of thousands of technology and business books

- Quickly search over 7000 books and certification guides
- Download whole books or chapters in PDF format, at no extra cost, to print or read on the go
- Copy and paste code
- Save up to 35% on O'Reilly print books
- **New!** Access mobile-friendly books directly from cell phones and mobile devices

Stay up-to-date on emerging topics before the books are published

- Get on-demand access to evolving manuscripts.
- Interact directly with authors of upcoming books

Explore thousands of hours of video on technology and design topics

- Learn from expert video tutorials
- Watch and replay recorded conference sessions

To try out Safari and the online edition of this book FREE for 45 days, go to **www.oreilly.com/go/safarienabled** and enter the coupon code RTHXZAA. To see the complete Safari Library, visit safari.oreilly.com.

Spreading the knowledge of innovators safari.oreilly.com

CPSIA information can be obtained at www.ICGtesting.com
Printed in the USA
BVOW060450191211

278569BV00005B/3/P